The Cambridge Nature Study Series

General Editor: HUGH RICHARDSON, M.A.

THE STORY OF
OUR TREES

Twelve-year old Spruce Plantation in Röken, Norway
(planted by school children)

THE STORY OF OUR TREES

IN TWENTY-FOUR LESSONS

BY

MARGARET M. GREGSON, B.A.

Cambridge :

at the University Press

1912

CAMBRIDGE
UNIVERSITY PRESS

University Printing House, Cambridge CB2 8BS, United Kingdom

Published in the United States of America by Cambridge University Press, New York

Cambridge University Press is part of the University of Cambridge.

It furthers the University's mission by disseminating knowledge in the pursuit of education, learning and research at the highest international levels of excellence.

www.cambridge.org
Information on this title: www.cambridge.org/9781107662780

© Cambridge University Press 1912

First published 1912
First paperback edition 2013

A catalogue record for this publication is available from the British Library

ISBN 978-1-107-66278-0 Paperback

PREFACE

THIS book is to help children to study Nature, not to put book study instead of Nature Study. The object of the book is to direct and stimulate both observation and reflection.

Many teachers have lately found in trees a convenient topic for lessons, one that can be pursued in town or country, summer or winter, indoors or out-of-doors. There may be other teachers who would gladly give their pupils some training in Nature Study, although they themselves do not pretend to have special knowledge. Such an acknowledgment of ignorance is itself a qualification for one who, an enquirer himself, would lead others to enquire.

The use of this book will be found greatly simplified by its arrangement into 24 lessons, each complete with its own practical work. In setting additional questions, it should be remembered that children are apt to answer observation questions correctly without using their brains more than they can help. The questions therefore, should be framed so as to evoke thought as well as observation.

Great pains have been taken in the revision of these chapters to fit them into the natural cycle of the seasons, which is now widely accepted as the best order for nature lessons, and is, indeed, the compulsory order

if free use is to be made of fresh material. The supplies
necessary for each lesson are clearly indicated. But
the lessons have also been adapted (and this is a
much more difficult adaptation) to the school year and,
again, to the different years in use at different schools.
For Secondary schools beginning with new classes in
September, the book begins at Chapter I. But, if some
Primary schools rearrange classes at Easter, the lessons
may begin under the stimulus of the spring at
Chapter XIII, and afterwards pass from Chapter XIV
in summer to Chapter I at the beginning of autumn.
Or, if some other schools make promotions in January,
it will do equally well to start at Chapter VII and work
round the year to Chapter VI in December.

A special effort has been made to assign as much
work as possible to the winter months, and thus to
meet the difficulties of any school time-table which
allots a uniform number of hours per week to Nature
Study, hours insufficient to cope with the splendour
of the spring, hours harder to fill in the gloom of
November. In this way, it is possible to treat some
aspects of the many-sided interests aroused by trees
which would have been crowded out by lessons intended
only for the summer term.

The serious inquiries now being made into National
Afforestation show that we may have to enlist the
sympathy of boys and girls in the planting and protection of new woodlands. The festival of Arbor Day
might well be transplanted from America to the rural
schools of the British Isles. A chapter on Forestry and
tree planting will, therefore, be welcome. In the revision
of this we have had the assistance of Mr S. Burtt Meyer,
of York, a trained forester. To him, our thanks are

also due for arranging for the use of some beautiful Norwegian photographs one of which, Fig. 32, shows the schoolboys at work clearing the ground, another, the frontispiece, shows the girls of Röken standing in a plantation of about their own age, which had been planted by school children 12 years before.

Mrs Gregson is a former student of Newnham College, and studied in Cambridge under Prof. H. Marshall Ward, whose well-known volumes have supplied many of the illustrations used here. Her lessons are adapted to classes where the ages range from 14 to 12 or even younger, that is to the higher standards of elementary schools, to preparatory schools, to the lower forms of secondary schools and especially to those who are taught privately at home.

Mrs Gregson has shown in numerous examples that style of large, clear, accurate drawing which every pupil may be expected to attempt. Teachers requiring a 'key' may refer to the exquisite drawings in Ruskin's *Modern Painters*, and to the latest standards of fidelity, Henry Irving's photographs in *The Nature Book*.

We must not let the trees hide the wood from us. The study of the trees is only the beginning of the study of the wood. The wood means far more than its trees alone. There is the undergrowth of brambles or of bracken, the carpet of spring flowers, ferns, mosses, dead leaves or pine needles. Then there are the insects, the purple emperor butterflies aloft and the ringlets in the glades. There are beetles boring in the rotten wood; indeed, a whole book has been written (by Mr Gillander of Alnwick) on *Forest Entomology*. After the insects come the birds—the tits, the gold-crest, the treecreeper, the woodpecker; and the owls and woodcock haunt the wood as well. The study of all this wealth of life is no idle or frivolous byepath, it is the essence of Nature Study, the study of all the complex web of relationships in which all living things and lifeless forces are bound together. Here, these explorations are left for teacher and pupil to pursue as opportunity of time and circumstance may allow.

The study from manifold points of view of this interconnection of woodland life may assist us in the attempt to unravel the complexities of social life in human society. Perhaps the poet was hardly thinking of evolutionary ethics when he wrote

> One impulse from a vernal wood
> May teach you more of man,
> Of moral evil and of good,
> Than all the sages can.

None the less, these simple lessons on trees may be used as a thread on which to string still greater thoughts all round the circling year, whilst great new ideas are transforming our minds as silently as the seasons transform the woodland.

HUGH RICHARDSON.

12, St Mary's, York.
May, 1912.

CONTENTS

LIST OF ILLUSTRATIONS

Figs 2 and 32 are reproduced from the *Journal of the Board of Agriculture and Fisheries* (Aug. 1911) by permission of the Controller of His Majesty's Stationery Office; figs. 1 and 3 are from photos kindly supplied by the Ontario Govt., 163, Strand, W.C.; figs. 6 and 51 are from photos supplied by the British Botanical Association; figs. 49 and 62 are from Tansley's *Types of British Vegetation* (Camb. Univ. Press); figs. 5, 24, 30 and 56 are taken from the Cambridge County Geographies of *Worcestershire, Lanarkshire, Breconshire* and *Nottinghamshire* respectively; fig. 4 is from Seward's *Fossil Plants*, Vol. II. (Camb. Univ. Press); fig. 13 is from a block kindly lent by Messrs A. Gallenkamp and Co.; figs. 7, 11, 15, 16, 19, 20, 21, 22, 34, 42, 43, 52, 54, 55, 57, 59, 60, 61, 63 [(1) and (2)], 64, 65, 66, 67, 68, 69, 70, 73 and 74 are taken from H. Marshall Ward's *Trees* (5 volumes, Camb. Univ. Press); figs. 8, 9, 10, 12, 14, 17, 18, 23, 26, 28, 29, 31, 35, 36, 37, 38, 39, 41, 44, 45, 46, 47, 48, 50, 58 (lower), 63 [(3) and (4)], 71 and 72 are all from drawings and fig. 33 from a photograph by Mrs Gregson, fig. 50 being copied from a school drawing by G. W. Maw, Bootham School. Figs. 40, 53, 58 (upper) are from F. Darwin's *Elements of Botany* (Camb. Univ. Press). Fig. 27 is reproduced by kind permission of Sir William Schlich from his *Forestry in the United Kingdom*.

CHAPTER I

INTRODUCTORY

During your walks in the country you will very likely have seen men with saws and long ropes cutting down trees and carting them away. If you have stood and watched them you will have noticed the skilful way in which they arrange the ropes so that when the tree comes down it falls where they wish it to, and does no damage. You may also have wondered what will become of the tree and to what use it will be put when it is taken away.

When trees are cut down they are usually taken straight to a saw-mill. Here a great saw cuts the trunks up into the planks which are found to be the most convenient form in which to store quantities of wood or to send it to other places. Although nowadays iron has almost entirely taken its place for shipbuilding, wood is still put to such a number of different uses that we should soon not have a tree left in the country if we depended only on our native supplies. Large quantities of the wood used in England are therefore imported from more thickly wooded countries.

In Canada tree felling takes place on an enormous scale. When the trees are cut down they are cut across into great logs and are dragged through the forest by teams of horses until they reach the steep banks of a river. The logs are then let go, and slide down into the river with a mighty splash. Hundreds of these

Fig 1. A Log Chute, Canada.

logs, jostling each other over the rapids, are carried down by the river. Sometimes they get jammed together and can go no further until men, taking their lives in their hands, leap from one log to another breaking up the block. When they reach the saw-mill

Fig. 2. Checking timber on the Norwegian Government forest reserve.

they are sawn into planks and are then placed on ships for transport to other countries.

We have now in England only a few forests left, but if we could look back some million years we should see a very different state of things. In those days the country was covered with swampy forests in which ferns and other plants grew together so closely that the weaker ones were often choked out of existence in the struggle. When they fell to the ground they gradually became rotten and other dead plants and leaves fell upon them. Thus, by degrees, thick layers were formed. Very many years passed by, earthquakes and other more gradual changes altered the surface of the earth, so that at last these old layers found themselves right deep down in the earth and became a hard black substance. This is what you know as coal. It is now found in coal mines in the depths of the earth and is dug up to be burnt in our grates. Sometimes in a piece of coal you will see the mark of a fern leaf, or of a piece of stem (Fig. 4). These are almost the only traces left to tell us what those ancient forests once were. If you are ever on a peat moor, look carefully at the black peaty earth. You will see that, even some way below the surface, it is full of little roots and stems of plants. This peat shows a very early stage of decay through which coal very possibly went before reaching the form in which we know it.

After many ages had passed by, England became inhabited by men, and as these inhabitants became more civilized they began to cut down small tracts of forest and to cultivate the land. The larger and fiercer animals that used to range through the country were extermin-ated and the forests came to be looked upon as hunting

Fig. 3. A Log Raft, Canada.

grounds for the king and his nobles—as such they were carefully preserved and wherever this was not the case they were looked upon as unfailing supplies of fuel and soon disappeared.

As the power of the people grew greater and that of the Crown less, more and more forest land was reclaimed

Fig. 4. Fern Leaf in Coal shale.

and used for agriculture until only a few Royal forests were left.

It was a great many years before the full value of our English woodlands was recognised, but in recent years many small plantations have been made to provide shelter for crops and for game (pheasants, deer, etc.). Where timber is the chief motive for a plantation the ground should be closely planted with trees that grow tall and straight.

Apart from their utility for timber and for those health-giving properties of which we shall speak in another chapter, trees are valuable to a country for the beauty they give it. Have you ever noticed, as you

Fig. 5. Typical Midland scenery. The river Avon at Evesham.

travelled from one part of England to another, how the differences in the scenery through which you pass are almost entirely due to them? Suppose you have been staying on the coast of Norfolk and you are going back to your home in Devon. You have been on sandy hills, with here and there clumps of Scotch Pine that give the country a rather wild and lonely look. As soon as the Fens are reached the land becomes flat, with silvery willows lining the banks of the winding streams. Then, all through the Midlands, you have rich pasture land dotted over with elms, oaks and beeches. This is typical English scenery, and is to be seen nowhere else. In Devonshire the climate is so mild that you will see many trees that will generally only grow in warmer countries—this makes many people who visit this county for the first time say how much they are reminded of the South of France or of Italy.

If instead of going across England you travel up to the North you will perhaps see the Yorkshire or Scotch moors. Here the hills are covered with short purple heather, and, owing to the cold winds, no trees will grow. Until you began to notice it for yourself you would never believe how great a difference trees make to the country. A few years ago one of the magazines published an article illustrated with pictures showing side by side the London parks of which we are so proud and the same parks as they would look shorn of their trees. No one would have taken them for the same places. Hyde Park and Kensington Gardens looked flat uninteresting wastes with no landmarks but broad paths, artificial lakes, and Hyde Park Corner and the Marble Arch towering up at the ends.

The trees of which we have been speaking call up in the mind a picture of something very tall with a thick woody trunk and spreading branches on which green leaves are borne. The mere size is not, however, always a safe test of whether a plant is or is not to be called a tree, as it partly depends on the conditions under which the plant grows. You are accustomed to think of a cabbage as a plant a foot high and of an oak as a tall stout tree. This idea is perfectly true, yet, in the mild climate of Jersey are grown cabbage plants over six feet high, with stems so strong and hard that they are made into walking sticks. As for the oak, the Japanese

Fig. 6. Japanese Dwarf Trees.

have a way of so pinching back young trees and repeatedly repotting them, that although beautifully shaped they never grow big. In Japan you may see oaks and cypresses 100 years old and yet small enough to be put upon the dinner table (Fig. 6). Yet, however stunted in growth these trees may be, they still show a thick woody trunk, and it is this trunk which chiefly distinguishes trees from other growing plants.

CHAPTER II

FRUITS AND SEEDS

LESSON 1

Season. About first week in October.

Materials required for each pupil.

One broad bean which has been soaking 12—24 *hours in water and one unsoaked one. One acorn in its cup. One of each of the following fruits or seeds:— hazel nut, rose-hip, grain of wheat, Spanish chestnut, horse-chestnut.*

Nearly every kind of tree passes the first stage of its existence wrapped in a seed in a tiny, sometimes microscopic, form. A very long time ago, before flowers as we now know them existed at all, some of the only plants that made any attempt at flowering belonged to the pine family. Their seeds were placed directly on scaly leaves and were uncovered, as is the case with pine seeds to-day.—If you slip a knife between the scales of a ripe pine cone in the early summer you will find

two naked seeds growing on the inner side of each
scale (Fig. 7).—As the years go on all living things tend
to improve themselves and to become more complicated,
and so, later on, some plants began to fold the seed-

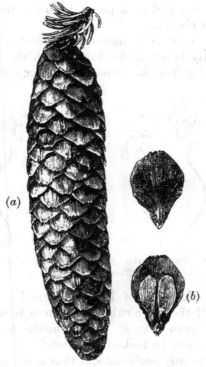

Fig. 7. (*a*) Pine Cone. (*b*) Upper and under surface of a scale from
the cone. The under surface bears two winged seeds.

bearing leaves over the seeds. At last all but the pines
did this. This is how it happens that flowering plants
belong to two great divisions; those with uncovered

seeds and those whose seeds are borne in a **seed-vessel**.
It is this seed-vessel which forms the **fruit**. Although
you have no doubt been accustomed to think that all
fruits must be juicy and sweet like those we eat, they
can equally well be hard and dry ; so that, botanically
speaking, a pea pod or a poppy head is a fruit quite as
much as is a cherry.

When a plant is in flower the young seed-vessel is
in or directly beneath the flower, with the rest of the
flower growing from it. When the flower fades it leaves

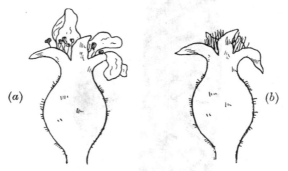

Fig. 8.   Young Apple.   (*a*) With remains of flower ;
(*b*) with flower rubbed away.

a few shrivelled bits of leaves fastened to the seed-
vessel, and if these are rubbed off there is still a mark
where they grew (Fig. 8). These marks are very important and must be looked for carefully, for sometimes,
when there is only one seed and that is wrapped tightly
round by the seed-vessel, it is only by finding these
marks that you can be sure you are looking at a fruit
and not at a seed.

The number of seeds produced by a single flower
varies greatly in different kinds of plants.   If the seed

is small and likely to be easily damaged there is generally a good deal of it there to make up for what is lost. For instance the common henbane produces about 10,000 seeds on one plant every year, and it has been calculated that if every one of these seeds became a plant there would be enough henbane to cover the earth's surface in three years. When the seed germinates the young plant is at first unable to draw its own food from the earth and the air as it will do later on. To make up for this some food which can be easily absorbed is always put ready for it in the seed. Our most nourishing vegetable foods, such as all kinds of corn, beans, and nuts, to mention only a few of the most important, are drawn from the supplies which have been laid by in seeds. Sometimes, as in wheat grains, the food lies all round the young plant, and sometimes it is all contained in the two fleshy seed leaves which, in the case of the bean for instance, fill up the whole seed. From these **cotyledons**, as they are called, the food is drawn as soon as the plant begins to grow. The whole is covered with a strong skin which protects the young plant, while still in the seed, from any sudden change of temperature. This skin does its work so well that some seeds are said to have been kept for nearly fifty years, and have then sprouted without being much the worse for it. The fact is that while a plant is in the seed it is in a sleeping state. As soon, however, as it is supplied with **moisture, air**, and **warmth**, all of which conditions are necessary for its growth, it begins to wake up. When this stage is once reached the seedling will die if any of those conditions are taken from it, even though it got on quite well without them while in the sleeping state.

Taking a broad bean, which is conveniently large, as a typical seed we will now look at it closely. If it has been soaking in water for 12—24 hours you will find that it has sucked up quite a lot of water. On comparing it with a bean that has not been soaked you will see that it is about twice the size of the other one, and that instead of being shrivelled up and hard it is quite plump with the water it has sucked up. Do you see the black scar on one side? this is where a stalk fastened it into the pod. Now squeeze the bean gently. A drop of water oozes out of a small hole near one end of the scar. Notice exactly where the hole is. Although it exists in every seed it is only in a big seed like the bean that you can see it without the help of a lens. It is through this hole that water and air will reach the germinating plant. When the young root begins to grow its tip presses right against this hole. As it grows it splits the skin round the hole until there is room for it to come through.

PRACTICAL WORK.

1. Draw (exactly four times its natural size) a broad bean which has been soaked. Mark the scar and the hole through which the root will grow. Draw an unsoaked bean to the same scale.

2. Take the skin off the soaked bean. Cut one of the fleshy halves carefully off, so as not to injure the young plant. Draw half the bean (enlarged four times) in the same position as last time, showing the young plant. Mark the larger end 'root,' and the smaller 'shoot.' Notice that the root end points exactly where the hole in the skin was.

3. Draw an acorn twice natural size, (i) in its cup, (ii) out of its cup.

4. Scrape part of the shell off. Describe in writing each different layer you find.

5. Take all the shell off. Separate the two halves of the seed. Draw half the seed with young plant four times natural size, i.e. × 4.

6. Draw the following fruits and seeds :—hazel nut, rose-hip, grain of wheat, Spanish chestnut, horse-chestnut. Make each drawing at least 2 in. long and write the scale at the side of each (thus, × 4). Say in each case whether it is a fruit or a seed, giving reasons.

HOME WORK.

Cut transversely through the unripe seed-vessel of any flowering plant and draw the cross-section about six inches in size as in Fig. 9.

Fig. 9. Seed-vessel of Narcissus cut across transversely.

LESSON 2.   (A WALK)

Season.   Second week in October.

Fruits are divided into two main classes :—fleshy fruits, which include wild berries as well as all the sweet kinds of fruit we eat, and dry fruits. Each of these kinds of fruit is sub-divided according as it contains many seeds or only one. For our two typical fruits of the fleshy class we will take a grape and a plum. We

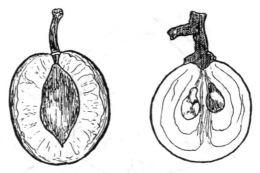

Fig. 10.   A Plum and a Grape cut through longitudinally to show the difference between a drupe and a berry.

find that the grape consists of a thin skin enclosing the sweet juicy pulp (Fig. 10). In the pulp are embedded several small hard seeds. The plum is like it as far as the skin and pulp go, but instead of several seeds there is one hard stone. The shell of the stone is really part of the fruit, and the soft kernel inside is the seed. A stoned fruit like this is called a **drupe**, while grapes and other similar many-seeded fruits are true **berries.**

You will understand more clearly the different types of seed-vessels by looking at this table :—

Fruit (=seed + seed-vessel)

Fleshy fruit		Dry fruit	
Many-seeded (Berry)	One-seeded (Drupe)	Many-seeded	One-seeded (Nut)

Opening by splitting   Opening by pores   Opening by lid

We now go on to the right-hand branch of the table. From it you see that, like the fleshy fruits, the dry fruits may be either many- or one-seeded. If a dry fruit is

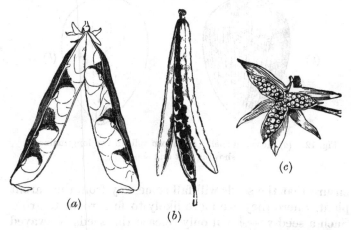

Fig. 11. Three types of seed-vessel : (*a*) Sweet Pea, (*b*) Wallflower, (*c*) Pansy.

many-seeded it must have some arrangement by which it releases the seeds when they are ripe, for they are often so small and so numerous that if they all fell to the earth together in their seed-vessel they would smother each other when they began to grow. Very frequently as soon as the seed-vessel is really dried up it splits open, in some cases with considerable violence, and scatters the fruit on the ground. Seed-vessels that split open are found in many different forms, some of which are seen in the accompanying sketch (Fig. 11). The pimpernel's seed-vessel and other similar ones have a lid which opens when the seeds are ripe. In a third class are those seed-vessels which like that of the poppy do not open at all, but are provided with a row of small holes near the top. The object of this arrangement is to

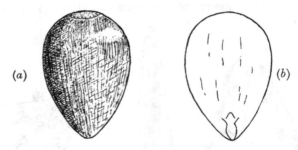

(a)   (b)

Fig. 12.   (a) Acorn in shell ; (b) Acorn split in two longitudinally, showing root and shoot.

ensure that the seeds will fall some way from the parent plant, where they are more likely to find room to grow. Such a seed-vessel will only release the seeds if swayed by a sudden knock or gust of wind, and this same blow will throw the seeds some little distance. Berries, although many-seeded, do not need any special arrangement for separating the seeds, as the birds that feed on them tear the fruit to pieces with their beaks.

The second class of dry fruits, the one-seeded ones, fall to the ground whole with their seeds inside them.

When the seed germinates the outer skin of the fruit either splits or rots away and lets the young seedling come through. The acorn which you drew last lesson is a good example of a dry one-seeded fruit. There is a tough skin (in the case of hazel nuts it is a shell) outside. The kernel lies inside this, and right in the middle is the young plant (Fig. 12). Many large and small fruits belong to this type, often called **nuts**, and, as you found in your practical work the other day, you have often to look at them very carefully before you can tell whether they are fruits or seeds.

At this time of year if you look under an oak tree you will find many acorns scattered about. Out of all the number that you see many will come to a violent end and will be eaten by squirrels or mice or rooks; but a few will survive the winter, and, if they have fallen on suitable ground will grow into little oak trees. Choosing out healthy ones take several of these acorns home, as we want to watch the first steps they go through as they become trees.

PRACTICAL WORK.

You will be taken for a walk. Collect as many different kinds of fruits as you can, paying special attention to those from trees. Find out the name of each from your teacher. Where the seed-vessel seems to open in a curious way you should try to get an unopened one and an opened one.

HOME WORK.

Draw the fruits you collected in your walk. Each drawing is to be about 2 in. across. Print by the side of each its name, class of fruit, and number of times you have magnified it.

LESSON 3

Season. Third week in October.

Materials required for each pupil.

One of each of the following fruits:—snapdragon, shepherd's purse, (columbine), lychnis, (buttercup), lucerne, beech nut, groundsel, wych elm, agrimony, blackberry, herb bennet, sycamore. The fruits in brackets ripen in the summer, dried ones should therefore be used if available, if not, they may be omitted. Yew twig with berries.

When seeds are ripe and ready to sow themselves it is important that they should not always be merely scattered on the ground under the tree or plant that bore them. If this happened some tracts of ground would be so crowded with the young of one kind of plant that they would have little chance of living, while a few miles away this same plant would be unknown. Since seeds have no power in themselves of moving any distance they make use of several outside agencies by means of which they are distributed over the earth's surface.

The most important of these agents are :—(1) birds and animals, (2) water, (3) wind. The many different sizes, shapes, and colours shown by fruits and seeds vary according to which agent is going to be used.

The various brightly coloured fruits we know depend for their distribution on **birds** which like their sweet flesh and eat them greedily, as those of you who have cherry trees in your gardens know to your cost. As long as the seeds are soft and unripe and the fruit hard it remains the colour of the leaves and passes unnoticed by the birds; but as the seed ripens the surrounding fruit takes on a sweet taste together with a bright red, black, or white skin which makes it easily seen from a distance. The seed itself either drops accidentally from the bird's beak or, being protected either by a stone or by a slippery skin (as in apple pips), is swallowed and passes uninjured through the bird. This is how it happens that wild roses and other bushes are sometimes seen growing half-way down the face of a

Fig. 13.   Coco-nut, showing the light fibrous husk in which it grows.

precipice and in other similar places that are inaccessible to anything but birds. Small animals like squirrels and field mice are very fond of nuts. As they run to and fro with them, adding to their winter store, many will be dropped and forgotten and will in time become trees.

Every time you go for a scrambling country walk you probably yourselves act as unconscious carriers of seed. Have you not often when you got home picked quantities of burrs from your clothes? You very likely brushed off others in the grass after carrying them some way. Sheep pick up so many of these burrs that the commercial value of wool is sometimes seriously injured by it. Special machinery is used for getting rid of them. The tiresome hooks with which they are covered are the provision they have made to force animals to help in their distribution.

Some large tropical seeds make use of **water** as their carrying agent. The best instance of this is seen

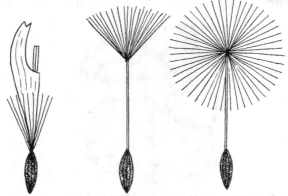

Fig. 14.   Three stages in the development of a winged Dandelion fruit.

in the coco-nut. The coco-nut palm grows on the shore of small coral islands, so that when the nuts fall from the trees they are often washed away by the sea. They are, however, wrapped in a boat-shaped husk that is tough but very light (Fig. 13); and in this they float securely until they are perhaps washed up on the shore of another island, where the damp and sun together soon make them germinate. This is why on newly formed coral islands the coco-nut palm is often the first form of vegetation to make its appearance.

The most common agent of all for scattering seeds is the **wind**. Many tree fruits have their skin produced into a thin sail-like portion which catches the wind without adding much to their weight. When ripe fruits of this kind are blown from the tree by the autumn gales, instead of falling straight to the ground they are carried away. All feathery fruits (you would probably like to call them seeds, but both fruit and seed are really there) like those of the wild clematis and the dandelion are again specially adapted for distribution by the wind (Fig. 14). Weeds with winged fruit are those that we find hardest to keep down in our gardens, for if only one head of fruit is allowed to ripen the air is soon full of seed and carries it everywhere.

PRACTICAL WORK.

1. Write down six fruits or seeds (not fleshy fruits or those mentioned in chapter or questions) used as food by men.

2. Arrange the following seed-vessels on a piece of paper and write under each one how you think the seeds are scattered :— snapdragon, shepherd's purse, (columbine), lychnis, (buttercup), lucerne.

3. Write a short essay (not more than one page) on the signs of autumn you noticed in your walk last week.

4. Draw the following fruits × 3 and write underneath each one for what distribution it is adapted, giving reasons :—beech nut, groundsel, wych elm, agrimony, blackberry, herb bennet.

5. Look at the berries on your yew twig. What is there curious about the position of the seed in the berry ? Is it inside the berry ? Make a drawing to explain what you see.

6. Throw a sycamore fruit high into the air, if possible out of doors in a wind, or let it fall from an upper window. Describe in writing how it falls to the ground. Draw the fruit × 3. Cut it in half parallel to the wing and again draw it × 3, showing young plant

HOME WORK.

Which of the following do you call fruits and why :—tomato, rhubarb, potato, hazel nut, acorn, artichoke, Brussels sprout, cauliflower, vegetable marrow, beetroot, ginger, lemon peel, raisin, almond, angelica, French bean, banana, onion, walnut, winter cherry, turnip, fig ? Arrange your answers in three columns, writing in the first the name, *e.g.* tomato, in the second 'yes' or 'no,' *i.e.* it is, or is not a fruit, in the third your reasons for this answer.

CHAPTER III

LEAF CASTING. (A WALK)

Season. Last week in October.

Materials required for each pupil.

Two acorns and two kidney beans.

When summer merges into autumn and the days grow shorter the leaves begin to change colour. The first frosty night often brings this change, and when morning comes the green of the leaves is found to be touched with yellow. During the next few weeks many splendid tints of gold and red will add beauty to our woods before the leaves finally fall from the trees. The reason for so gradual a change is that the leaf is joined by a living stalk to the stem, and if the leaves were all suddenly cut from the tree they would leave a lot of open wounds behind. This is how the trees prevent it: all through the summer moisture has been passing up from the earth into the leaves, but as the weather grows colder the trees live less actively and this drawing up of water gradually ceases. While this change is taking place, a thin layer of what is really dry **cork** begins to

grow from each side of the leaf's stalk where it. was
joined to the main stem. As the leaf dries up the green
colouring matter which gives it its bright colour and
is called **chlorophyll** breaks up and disappears, its
place being taken by various yellow and red substances
according to the kind of leaf. At last the cork has grown
right across the leaf stalk except just where the veins

Fig. 15. Falling Horse-chestnut leaves.

pass into the leaf. At this stage the leaf is seen to be
hanging on by a few threads only. Then the cork layer
joins right across the middle and the leaf drops off. In
this way the wound on the stem is never exposed to the
cold air as would happen if you tore a leaf off before
it was ready to fall. Of course leaves are not always
allowed to hang peacefully on the boughs until they

drop off of their own accord. The autumn winds often blow them off while they are still green. Thus it happens that you will find leaves showing every stage of autumn colouring under one tree. On the other hand some kinds of trees keep their dead leaves on the twigs to act as a protection to the young buds and do not shed them till the next spring when the weather is milder and the buds are able to take care of themselves.

Fig. 16.   Old Scotch Pine, showing stumps of branches.

Many trees cast in this same way not only leaves but twigs. You have very likely noticed how the ground round some trees at this time of year is strewn with twigs and small branches, and have thought the wind was to blame. This may be true in part, but it is not due to the wind alone. This **branch casting** is a tree's natural way of getting rid of any twigs which, because they are crowded or badly placed, it does not want. Scotch pines go even further than this, and, as they grow old, cast their branches all the way up the trunk (Fig. 16). On any old pine you will see the stumps left on the sides of the trunk where the branches used to grow. This will happen most where trees are crowded together in woods. Foresters know this and depend upon this process for freeing the trunk from small useless twigs and branches, or as they call it "cleaning the bole."

All through the months of October and November the falling leaves are a great trial to the owners of trimly kept gardens and parks, and the gardeners are kept very busy clearing them away. If you remember the great heaps that they collect you will wonder what becomes of the leaves in a wood, where there are no gardeners. Do you know what happens to the garden heaps? They are all carefully piled together in some sheltered corner where they cannot blow away and are there left. If they were dry to begin with, rain and dew will soon damp them, and when they have lain there for a few weeks they will turn dark and soft and will begin to rot away. This rotting is really a very slow kind of burning, and if you put your hand into the middle of the heap you will find it quite warm there. As it goes on the dead leaves break up until all that is left of them is a much smaller heap of dark brown mould. Although it takes up much less room than the leaves it contains all the rich plant food that was in them. Because it holds much nourishment in a little space this leaf mould is very valuable for gardening and is mixed with the earth used for potting plants.

This process is something like what goes on in the woods. The fallen leaves cover the ground thickly and, as they lie there exposed to the damp and to the air, they rot away and leave in their place a thin layer of rich dark soil. In a wild wood this **humus**, as the leafy top soil of the woodlands is called, goes on collecting year after year and provides the trees with just the food they want. If, however, the dead leaves in a wood are collected and taken away, the trees will be half starved, for when they lose their leaves they will get nothing back from them as they would have done if left alone.

PRACTICAL WORK.

You will be taken for a walk to study leaf falling. Notice the following points : —

1. Are some kinds of trees barer than others ? If so, which ? What kinds of trees show least change in their leaves ?

2. Do you see any trees whose leaves though quite brown are not falling ? If so, which ?

3. Can you find on any tree a twig or branch which has been partly broken off in the summer and has died ? What is happening to the leaves on such a branch ?

4. Collect leaves from under any one tree to show four to six stages of change of colour.

HOME WORK.

1. Draw the outlines of your leaves in pencil (natural size) and paint them to show the different stages of colouring.

2. Take home with you from your lesson two acorns and two scarlet runner beans. Place them in a saucer half full of water. Look at them at about the same time every day and write down on a piece of paper any change you notice in them and the day on which you saw it. You will want these notes for your home work next week. Also draw in pencil each day an outline sketch showing these changes.

CHAPTER IV

THE GROWTH OF SEEDS

Season. First week in November.

Materials required for each pupil.

The saucer containing the beans and acorns must be brought from home. In addition each child will want two acorn glasses or test tubes fitted with paper funnels and labels, and two of each of the following seeds: sweet pea, horse-chestnut, wheat[1], sycamore, pine. A warm room, greenhouse, or cupboard warmed by hot pipes. (If the seeds are grown entirely without heat germination will be found tediously slow.)

If you have been looking every day at the acorns that you put to soak a week ago you will have seen that after a few days three even cracks appeared at the pointed end. To-day a small white point is beginning to force its way between the cracks. This point is the tip of the **root** of the growing oak tree. If you split open one of your acorns at this stage, and compare it with what you saw last time you drew the inside of an acorn, you will find that the other end of the little plant, which is called the **shoot**, and will in time grow into the leaves and branches, has hardly grown at all yet. Although the beans were put to soak on the same day as the acorns they are already showing quite long

[1] Before purchasing wheat from seedsman, brewer, or flour miller, it is well to enquire whether it is likely to grow. Some wheat which is on sale as foodstuff has been killed so that it will not do so.

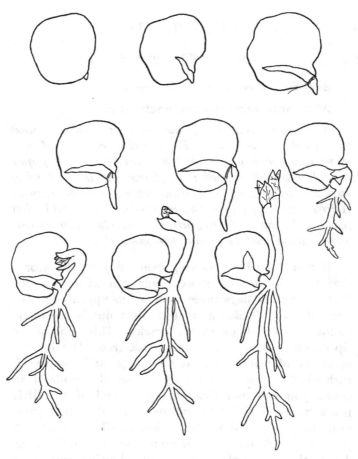

Fig. 17. Stages in the germination of a Broad Bean grown in a test tube,
showing bean after 2, 3, 7, 9, 12, 14, 15, 16, 17, days.

roots. They will go through their short lives in one season and will show us quite soon stages that it will take the acorns months to reach, and so we are going to grow the two kinds of seeds side by side.

We now want to be able to watch them growing, and for this we must put them in glasses so that the roots will have room to grow, but will not be hidden as they would be if we planted the seeds in the earth. The best kind for this purpose are the acorn glasses which can be bought at most large china shops for two or three pence. If you cannot get one of these you will be able to manage with a specimen glass or a test tube if you arrange a paper funnel at the top so that, although the root can get through, the seed cannot slip in. The glasses should be nearly filled with water, so that the tips of the roots just touch it when the seeds are put in roots downwards.

If instead of putting them this way up you allowed your seeds to lie on their sides you

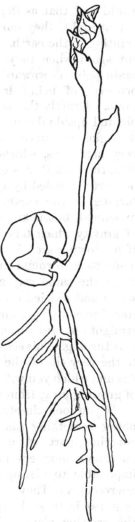

Fig. 18.   Broad Bean plant
21 days old.

would find that as the roots grew, instead of growing
out straight, they curved right round with their tips
pointing to the earth. They soon straighten themselves
out again when they are put the other way up, and
again point downwards. All roots possess this strange
property of being drawn towards the centre of the
earth. Exactly the same thing happens if an acorn is
planted upside down. As soon as the root shows itself
it begins to curve round towards the earth. This is
very useful, as, whichever way up an acorn is when it
falls, the root always grows down into the ground. This
property is called by a long name, **geotropism**, meaning
turning to the earth. It has never been altogether
explained, but is known to be connected with the force
of gravity which tends to draw everything to the earth.
If a growing seed is placed on a revolving table and is
made to turn round and round by clockwork, then, in
much the same way as a stone tied to the end of a
string and whirled round your head pulls outwards, the
root instead of growing towards the ground will grow
straight out, away from the centre of the circle in which
it is turning. This sensitive behaviour is found to belong
to the very tip of the root. Other properties which it
has and which you will see are very useful to it are those
of growing away from the light and towards moisture.

If you look closely at the root of your bean you will
notice something that looks like a thin layer of gum
covering part of it. This layer really consists of
enormous numbers of tiny hairs, so tiny that it is
impossible to distinguish them without the help of a
microscope. They are called **root hairs**, and their
business is to suck up the water and to pass it on to
the growing plant. They are so small that they grow

into all the crevices in the earth, and so thin skinned that the water can easily pass into them. They stick to particles of soil and draw from them all the moisture that is wanted. These hairs only live a few days, then they die down and others spring up nearer the tip to take their place. This process always goes on, even in an old tree, for the tree must have water, and it is only through the root hairs that it can obtain it.

After your bean has been growing for about another week its skin will begin to split. Then the stalk joining root and shoot grows longer and finally pulls the shoot out. This shoot has exactly opposite properties to the root. It grows towards the light and directly away from the earth, so that, in whatever position you place it, it grows straight up, just as the root grows straight down. This property is very important to heavy trees, for if it did not exist they might grow crooked, and then their boughs and leaves would weigh them down and they would fall by their own weight, just as a crooked pillar would.

When the shoot has come out, the split outer skin gradually comes off and the two cotyledons separate. In the bean and acorn they remain underground and, having fed the young plant as long as it needed it, rot away. In many trees, however, they come to the top of the ground with the shoot, and, becoming green, appear as the first two leaves. This is what happens in the sycamore and in all other young plants whose first two leaves are quite a different shape from any of the later ones. Two is not the only number of cotyledons a seedling can have. It may have nve to ten, like the pines, or it may have only one. Plants are called **poly-cotyledons**, **di-cotyledons**, or **mono-cotyledons**

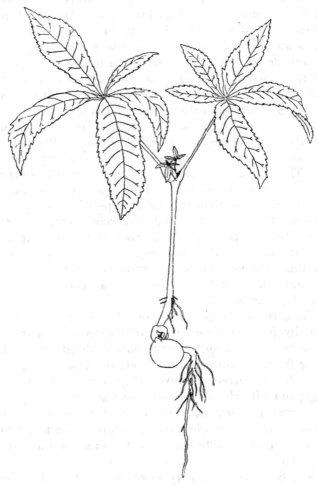

Fig. 19. Horse-chestnut seedling.

according to the number they have. To this last class belong the palms and all our English flowering plants in whose leaves the veins are arranged in lines instead of making a network. By pulling a young plant out of its seed coat you can generally see for yourself to which of these three classes it belongs.

PRACTICAL WORK.

1. Write down, one under the other, all the dates since your last lesson. With the help of the notes you made write by the side of each date any change you noticed in your seeds.

2. Outline in ink the pencil sketches you made of your seeds each day, and add the dates.

3. Fit your acorns and beans into the tops of the test tubes and pour in enough water to cover the ends of the roots. Write on the label your name and the date when the seeds were first put to soak (one week ago).

4. Find out by careful dissection whether each of the seeds given you is a mono-cotyledon, di-cotyledon, or poly-cotyledon.

Leave your dissected seeds on a sheet of paper for inspection, and write by the side of each one its name and the number of cotyledons.

HOME WORK.

Take your two test tubes home and, as before, write down and draw each day any change in them.

CHAPTER V

WINTER BUDS AND TREE FORMS

Season. Second week in November.

Materials required for each pupil.

Twigs of the following trees, each showing about six buds and leaf scars :—horse-chestnut, ash, beech, sycamore, elm, willow.

When the trees have lost their leaves their active life ceases for a time and they sleep until the warmer weather comes. All the new leaves and flowers that will be wanted in the spring are therefore made beforehand during the autumn and, in the form of buds, are wrapped up tightly to protect them from the winter's cold winds and frosts. These young buds are placed where they are least likely to get knocked off, namely in the corners where the leaf stem grows out from the twig. This corner is called the **axil** of a leaf and although sometimes, as in the elm, the buds may be a little shifted to one side, you will nearly always find a bud in the leaf axil and nowhere else (Fig. 20).

When dead leaves drop off in the autumn they leave behind them a mark where they were fastened on to the twig. In some kinds of trees this is small and hard to see, but in the horse-chestnut it is a big **scar** something like a horse's hoof in shape with marks like nails

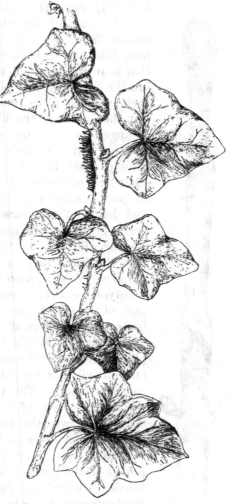

Fig. 20. Climbing shoot of Ivy, showing buds in the axils of the leaves.

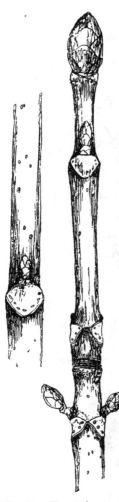

Fig. 21. Horse-chestnut twig, showing leaf scars.

in it (Fig. 21). These nail marks are the ends of the fibres that passed from the twig carrying food into the leaf. Other trees that show well-marked leaf scars are the ash, oak, and plane. In each case the shape is different, but they all show the fibre ends arranged in a definite shape on the scar. Anyone who has studied trees can say in a minute from what kind of tree any twig comes by just looking at the leaf scars and buds. You yourselves will get a great deal of interest out of your winter walks by noticing how the buds vary in size and shape and you will soon recognise when you see them the round black buds of the ash, the pointed beech buds, and a number of others.

If you look for the buds on a plane tree before the leaves drop off you will not be able to find them anywhere. This is because the swollen base of the leaf is drawn out over the bud into a sheath which hides and shelters it completely. When the leaf is pulled off the bud is found underneath. The leaf scar right round the bud shows you where the leaf once grew (Fig. 22).

Buds vary not only in their size and shape but in their arrangement on the twigs. Thus on a horse-chestnut twig they grow **opposite** each other, while in the beech they are placed **alternately.** In both cases they are arranged quite regularly along the stem just as the leaves (as shown by the leaf scars they left behind) grew in the summer. These buds grow into twigs, and the twigs, in time, into branches, and so it follows that in some kinds of trees the branches are opposite, while in others they are alternate. This is

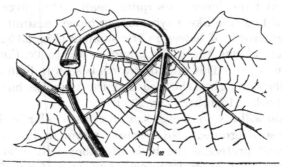

Fig. 22. Leaf of Plane, with bud.

one of the differences that should strike you when looking at such trees as the sycamore, horse-chestnut, fir, on the one hand, and the oak or elm on the other. Notice also how the way in which the main branches grow out from the trunk depends on the kind of tree. In the ash they grow up fairly close to the trunk, while in the oak and fir the branches sweep straight out almost horizontally.

There is just one other point about the growth of trees that I want you to notice for yourselves while the shape of the bare boughs can easily be seen. Is it on

the whole the buds placed on the outer side or on the
inner side of the main boughs that have grown into the
largest branches? You will find that in some cases one
thing has happened and in some the other, but the same
is always met with in the same kind of tree. You will
understand what I mean if you take a simple outline of
a tree like the figure opposite, with unforked branches
growing opposite each other, and draw in yourselves
the smaller branches and twigs:—In the first case make
the branches growing outwards into thick many forked
ones, and the inner ones quite small. This gives you
a spreading tree like a sycamore or horse-chestnut, with
a large crown of leaves in the summer (Fig. 23 *b*). In
the second one the inner branches are to be the important ones. This kind of growth is seen in lime trees
and lilacs and gives the idea that the branches have all
been pushed together (Fig. 23 *c*).

You will say you never saw trees as regular in shape
as these diagrams. No, because although I told you
how well protected the buds are, they still often get
knocked off or eaten by insects, and, even if a leaf bud
is not altogether destroyed, a small accident to it will
mean that later on a whole branch will be stunted in
growth. You will sometimes see a small Christmas
tree (spruce fir) or a young maple growing beautifully
regular, but as trees grow older each individual begins
to have a shape of its own. For this reason it is much
easier to notice different patterns of branching on young
trees.

Although trees will suffer if they do not get plenty
of air, yet if this air takes the form of continual wind
they may easily have too much of it. When they grow
along the sea coast, or in any other exposed position

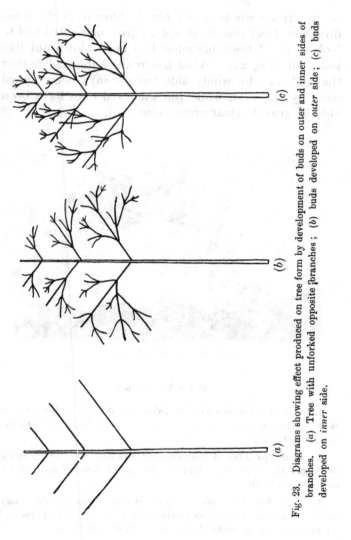

Fig. 23. Diagrams showing effect produced on tree form by development of buds on outer and inner sides of branches. (*a*) Tree with unforked opposite branches; (*b*) buds developed on *outer* side; (*c*) buds developed on *inner* side.

where strong winds nearly always blow from the same direction, trees are often apt to grow one sided and to look as if all their branches had been blown out like long hair (Fig. 24). What has really happened is that the buds on the windy side have always been nipped back by the cold, while the sheltered ones have been able to grow to their proper size.

Fig. 24. Wind-blown tree.

PRACTICAL WORK.

1. Draw the leaf scars of the twigs provided × 6. Show in your sketches the bundles of veins in the scars, and write underneath the name of each twig.

2. Draw any three of the twigs life size with the buds growing on them. Draw one of the buds by the side of each × 6, and put the names underneath.

3. Draw two diagrams and show the branching of imaginary trees with the chief branches leaving the trunk at right angles and (*a*) buds developing inwards, (*b*) buds developing outwards.

4. Draw two diagrams to show the branching of imaginary trees with the chief branches drooping downwards and (*a*) buds developing inwards, (*b*) buds developing outwards.

HOME WORK.

Sketch from nature the forms of six trees in which you can recognise the branch pattern, showing in each case the trunks and branches. Put each sketch on a separate sheet of paper with the name underneath.

CHAPTER VI

TREE PLANTING AND TREE FELLING

LESSON 1. (A WALK)

Season. Third week in November.

Although when a country is prosperous and thickly inhabited its prosperity can, as a general rule, be seen in its abundant cornlands, it is important that there should be some woods scattered about among the cultivated land to act as a shelter for the crops as well as for the timber they yield. During the latter part of the nineteenth century it was at length realized that in England trees were being cut down far more rapidly than they were growing up, and that at this rate we should soon have hardly any trees left. Since then a little more care has been taken as old trees are felled to make plantations of young ones that will in time take their place. In order to do this successfully and to get the best possible yield of timber it is necessary to understand something of the conditions under which trees grow best. In making plantations the wants of different kinds of trees must be studied as those which grow well on the outskirts of a wood might not thrive in the middle. Again, in a **mixed wood** it is necessary to place **shade-bearing** trees like beeches and silver firs under such **light-loving** ones as the birch and oak, when it would not be possible to do the opposite. Some trees like the oak, elm and larch are far more particular as to where they will grow than is, for instance, the pine. When farmers began to take all the best land for growing corn the pine was often the only tree that would grow on the poor high-lying ground that was left. This is how it happens that in European forests the pine is commoner than any other tree. The science that tells you things of this kind and makes itself responsible for the preservation and care of woods is called **Forestry**.

Suppose now that you have a tract of land that has been allowed to run wild for some years, and that you have decided to turn it into a plantation, knowing that, although you will have to wait a long time before you get any timber and hence any return for your money, it will in time pay steadily.

The first thing to be done is to clear the ground, removing any bushes or coarse grass. You must then consider the nature of the soil and what trees are likely to do well there. If the earth only runs down a short distance, and there is rock or sand underneath, you will only be able to grow trees like the pine and beech whose roots grow near the surface. If the ground is low lying and inclined to be damp the ash or poplar would be suitable trees as they like plenty of water ; if on the other hand it is very dry and sandy it may be that

Fig. 25.   20-year old Fir trees.

nothing will grow but the pine. The oak does best on
a *clay* soil, and the beech on *chalk*. We will suppose
that your land consists of good fertile **loam** (that is
a mixture like most garden soil), and that it stands
rather high, so that it is exposed to the wind, but that
the rain drains away nicely. Your best plan will be to
first plant some hardy kind of tree that does not mind
the wind. When these have grown big enough to afford
some shelter you can grow some less hardy tree that

Fig. 26.   Planting a young Pine.   (The rich mould from the surface is
placed on one side of the pit and the lower soil on the other.)

would otherwise be unable to stand the exposed position.
The sheltering trees are called **nurses** and for them you
cannot do better than choose Scotch pines. They are
quick growers and will not mind the wind.

If you want to be very economical you may try
to rear these from seed; but you will not want a
great many of them and time is important, so we will
buy young plants at the nurseryman's. Trees about
one foot high will do nicely and will only cost a few

shillings per 100.   The time to plant them is either the
autumn or the early spring.   The planting must be done
carefully, as you do not want to stop their growth more
than can be helped.   Dig a square hole big enough to
take the roots quite comfortably.   Now with the richer
soil from near the top make a mound in the hole.   Place
the little tree on the mound so that the roots are nicely
spread over it, and will find the rich soil as soon as they
begin to grow.   Fill in the hole and stamp the earth
down firmly (Fig. 26).   This is the method of planting
used for all trees that, like the pine, have a number of
small roots.   The oak and other trees with a single large
root have to be treated differently.   The trunk of the
tree should be covered to just the same level as it was
before it was moved.   If the trees are 12 to 15 feet
apart it will be near enough as we shall want room in
between them for the others.   A small plot of land
forming a square of which each side measures 75 feet
would want six rows of pines with six trees in each row
(see diagram).

What are you going to choose for the other trees of
your plantation?   Each of our native trees is valued
for its own special properties; the beech and elm for
the hardness of their wood, the ash for its elasticity,
and the pine for its clear grain and freedom from knots.
Foremost among them stands the oak, whose wood is
found to combine these important qualities better than
that of any other tree.   If your soil is suitable for growing
oak trees you cannot do better than choose them.   You
may as well grow them from acorns to learn how this
is done, but as a matter of fact the oak is such a slow
growing tree that in actual practice you would probably
buy the young trees to save time.   The acorns should

be sown not in the plantation, but in another piece of
ground that has been chosen for a **nursery** because
of its sheltered position. You can collect the acorns
from a neighbouring wood and imitate nature as nearly
as possible by sowing them in the autumn soon after

Diagram of plantation.

they fall. The ground must first be well manured, dug
over and raked. The seeds are then sown in long lines
or **drills** and in the case of acorns should be covered to
a depth of about an inch. (Other kinds of seeds if small
and light are sown nearer the surface.) You will have

to put a fence round your seed ground to keep out rabbits, and set traps in case mice come for the acorns. After this, beyond attention to weeding and watering, nothing can be done but wait for your trees to grow.

PRACTICAL WORK.

You will be taken for a walk to a wood or plantation. Notice the following points :—

1. Is the wood made up of one kind of tree or more ? If there is more than one kind in what proportion do they grow ?

2. Do all the trees seem about the same size and age ? are there any seedling trees growing up ?

3. Is there any undergrowth ? Does it grow right under the trees ? of what does it consist ?

4. Examine the soil under the trees. Is it dark or light ? Does it contain sticks or dead leaves ?

HOME WORK.

Write a short description, not more than two pages, of what you saw in the wood.

Lesson 2

Season. Fourth week in November.

Materials required.

A visit to a timber yard or saw-mill should be arranged.

In about two years' time you will find that the oak seedlings are beginning to be crowded, and that, although they are not yet strong enough to be moved straight into the plantation, they want more room. They must therefore be transplanted wider apart, and for this you had better prepare a long trench as close to the nursery as possible. Now take each plant up very carefully, loosening the ground round the roots so as not to have to use any force in pulling it from the ground. If the weather is particularly dry and the trees have to be moved some distance it is best to dip the root of each seedling into wet mud as you take it from the ground. The mud will soon dry and form a protective coat. This precaution is not however usually necessary. Your trench should be made with one side going perpendicularly into the ground. Each seedling should be held against this steep side while the earth is filled into the trench and is firmly stamped down.

After two years more you may either transplant your oaks again, or they may be sturdy enough to go into the plantation. We will suppose that they are. How close together are you going to put them? This all depends on what sort of timber you want your trees to yield. In former days, when the curiously twisted boughs of oaks were wanted for the framework of keels of wooden ships, the most useful trees were those that had plenty of room for their side branches to grow. If they are grown closer together, as is generally the case now, the lower branches are gradually smothered, but the top of each tree grows as tall as possible in order to overtop the trees round it and so reach the light (Fig. 27). Your probable object is therefore to place them near enough together to form tall straight trunks without many branches, but not so near that they injure each other. The right distance of course depends on the size of the young trees when you plant them out, and varies from about 3 to 5 feet.

Instead of planting the oaks in pits as you did the pines you had better use a wedge-shaped spade made

Fig. 27.   100-year old Pine forest in Thüringia.

on purpose (Fig. 28). By driving it into the ground and working it backwards and forwards a slit is made into which the root is placed. The slit is filled up by driving the spade into the ground a little further on (Fig. 29). The reason you use this method is that, while the pine has a number of small roots, the oak has a single **tap root** which goes straight down into the ground. The trees may now be left to grow until they become inconveniently crowded.

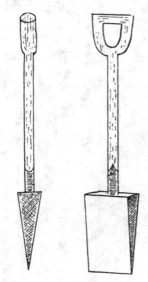

Fig. 28. Tool used in planting young Oaks and other similar trees.

In about five years' time you will probably find that the pines are quite overshadowing the oaks and must be removed. Trees may be felled with the axe or saw or by using both. In cases like this one, where the trees have no tap roots, it is sometimes possible to saw through the roots and so fell the tree. In this way the whole trunk can be used. Besides which the ground is left clear to plant new trees. Another method of tree felling that is more commonly used is the following :—The man who is to do it first decides on which side he wants a tree to fall. Then, when he has taken off some of the branches, he makes with an axe a deep notch on that side. He now takes a saw and, beginning on the opposite side, saws the trunk through. As there is no support where the notch has been cut, down comes the tree in that direction (Fig. 31 *a*). A third case is that in which the axe only is used, when a **V**-shaped notch is cut in one side and a deeper notch on the side on which the tree is to fall (Fig. 31 *b*).

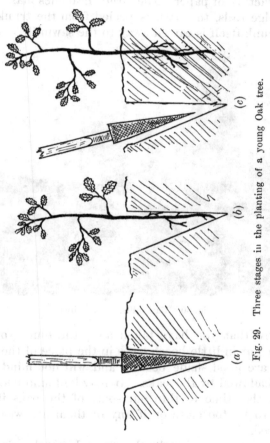

Fig. 29. Three stages in the planting of a young Oak tree.

When a tree has been felled, an operation for which the winter is the proper time, the next thing to be done is to sort the timber into various sizes. The smaller branches are cut up for firewood or, especially in the case of pines and firs, to be used as wood pulp in the manufacture of paper. The larger branches make props or hedge rails, the bark is peeled from the trunk, and the trunk itself is carted away to the sawmill.

Fig. 30.   Saw-mills, Breconshire.

Now that you have taken away the pines you had better put in little beech trees in the place of the pines. They are good shade bearers and will not mind being overshadowed by the oaks. It may be found necessary about this time to thin out some of the oaks if they seem to be too close or if any of them are weakly or diseased.

Some trees, especially the oak, ash, hazel and willow, have the power, if they are cut down fairly young, of sending up a number of new shoots from their stump. These shoots are valued in the oak for their bark, and in the ash and willow for basket making. They can be collected every few years, and others will spring up in their place. A wood that has been treated like this is called a **coppice**. If you want to turn your plantation into a coppice the oaks will be ready for felling when they are about 20 years old, and they must of course not have their roots taken up but should be cut down close to the ground.

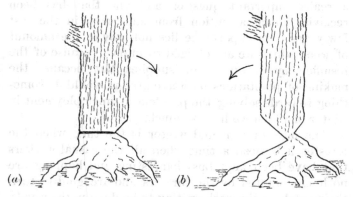

Fig. 31. Two methods of tree felling. (*a*) With axe and saw ; (*b*) with axe only.

In considering this plantation of yours I have made no mention of the many trials you may have had to fight against. Rabbits and other small animals often do a lot of mischief by nibbling bark from the young trees. Other dangers are storms, fires, and insect pests. We guarded against the first of these three by using a mixture of deep and shallow rooted trees, and fires are not likely to do much damage in the damp climate of

England except perhaps in the case of large forests of pine trees, which are specially inflammable because of the resin they contain. Insect pests are best prevented by planting two kinds of trees together and by keeping them in a healthy condition.

Your plantation has taken a long time to grow up, and the chances are that not yourselves but your children or grandchildren will get the full benefit of it. You may, however, comfort yourselves with the thought that this restocking of England with trees is a really important question and one that has been receiving much attention from statesmen in the last few years. Its importance lies not only in the amount of wood which we are obliged to import because of the insufficiency of our home supplies, but because the making of plantations on waste ground would do something towards solving the problem of unemployment in winter of which we hear so much.

The late autumn and winter is the time when the forester is busiest, a time when all the casual workers that have been kept busy harvesting and hopping are no longer wanted by the farmers and are glad to turn their hands to the work waiting to be done in the woods.

Many schools now keep an **Arbor day** for tree planting once a year. The children try to collect from friends the price of a tree, say 6*d.*, and as many as succeed in doing so set out on Arbor Day with spades and trees which they plant along a new street or in a piece of waste ground that has been chosen for the purpose. In this way some of our ugliest towns are being gradually beautified with green trees. The time chosen is either the beginning of November or the end of March.

Fig. 32. Norwegian School Children preparing land for tree planting.

PRACTICAL WORK.

If possible you will be taken for a visit to a timber yard or saw-mill. When there notice the various sizes into which timber is sawn, *i.e.* 'boards,' 'deals' and 'battens.' Notice also how the different kinds of timber vary as regards closeness of rings and freedom from knots. Take notes on what you see.

HOME WORK.

Take your oak seedlings from Chapter IV home with you and plant them in some piece of waste ground. Cut two twigs of elder or willow about a foot long and plant them in some corner where they will not be disturbed.

CHAPTER VII

EVERGREENS. (A WALK)

Season. First week in February.

Materials required.

(*a*) **For demonstration**, *small blocks of timber from various conifers (see Appendix).*

(*b*) **For each pupil** (*for home work*), *a leafy twig (about 6 inches long) of some known kind of Pine or Fir. Water colour paint box.*

The trees of which we spoke last term were for the most part **deciduous** trees, that is to say trees that lose their leaves before the winter. **Evergreens**, as their name implies, are trees whose leaves remain on the branches throughout the year. You must not imagine from this that the leaves last for ever, for each one in time reaches the limit of its life, usually after 4 or 5 years. Instead, however, of all falling off at once, leaving the tree bare, evergreen leaves are renewed by degrees, as they grow old.

In tropical countries most trees are evergreen, for there is no winter to interrupt growth, and although the trees may sometimes suffer from lack of rain, their tender leaves are never nipped by cold winds and frosts as they would be in Europe.   In North Africa and the southern parts of Europe we find what is known as a **sub-tropical** vegetation, where deciduous trees and pines grow side by side with the hardier palms.   Further north, where the winter grows colder and longer, life for the tropical plants becomes impossible.   Trees that keep their leaves throughout the year are still to be found, but they are very unlike the broad-leaved palms of the Tropics.   The evergreens of Northern Europe are nearly all **Conifers**, *i.e.* of the great cone-bearing tribe, and have a strong family likeness to each other. They are hardy trees and will thrive in mountainous districts where little else will grow.   The steep rocky slopes on the sides of Swiss mountains are generally thickly clad with pine trees whose pointed tips and gloomy colouring seem to suit the rugged character of the scenery.   On the outskirts of these woods may be seen quantities of small seedling trees springing up. This would not happen if the conditions under which they grow did not thoroughly suit them—conditions in which the ground beneath them is often frozen hard, while their branches are continually exposed to high winds and to sudden changes of temperature.   The leaves are often covered with snow or dripping with moisture in the early morning, and a few hours later they may be bathed in bright sunshine.   This variable weather is typical of the Alps and of other great altitudes not only in the real winter months but during the greater part of the year, and would soon kill the leaves of ordinary trees. It is, however, exactly for these trying circumstances that the leaves of the fir

Fig. 33. Swiss Pine trees in September. (This gives some idea of the wintry conditions trees at a great height have to endure for most of the year.)

family are adapted. They are small and narrow so that there is very little surface to be chilled or torn by the wind. They are also tough and shiny in order that frost may not hurt them and snow will slip easily off their polished surfaces.

If, as is the case with many trees, the branches of
pines formed a sharp upward angle with the trunk, snow
might slip off the leaves into this fork, gradually
collecting there, until there was a danger of its weight
tearing the branch away from the trunk. The branches,
however, instead of doing this, grow out from the trunk
almost horizontally, and even, in the case of the older
ones, sweep downwards, so that snow soon slides from
the branches straight to the ground.

When you were learning about planting trees last
term I told you that pines, instead of, like the oak,
having a single main root growing deep into the ground,
have a number of smaller ones which creep along
comparatively close to the surface. These enable the
tree to grow and prosper where quite a shallow layer of
soil covers the rocks, and where a tree with a deep root
would be unable to get any hold of the ground.

The spruce fir, larch, and other less familiar conifers
have been imported into Great Britain and now form
plantations all over the country. The Scotch pine is,
however, our only native representative of the family.
In former days forests of this tree covered large tracts
of land in Scotland and Ireland, but, although when
planted anywhere it often sows itself, it is only in parts
of the Highlands that it is now to be found growing
wild in any quantity. If you look at a twig of Scotch
pine you will see that the leaves are very long and
needle shaped, with a single unbranched vein growing
down the middle. (This single vein is characteristic of
all the conifers. They do not need much liquid food,
and if they did they would not be able to get it out of
the frozen or sandy ground in which they often grow.)
The leaves grow in pairs, each pair being bound together

this is to give an old tree an umbrella-like shape which
is quite different from what it was in its younger days.
A young Scotch pine is very regular in its growth.
Each year it adds a new circle of branches to its trunk
and one to each of the side branches. Thus, until it
begins to cast its lower branches, you can tell the age
of one of these trees by counting the number of circles,
or **whorls** as they are called, of branches growing out
from the trunk, and adding on one for the first year
before the seedling formed side branches. The accompanying diagrams will make my meaning clearer
(Fig. 35). For the sake of clearness I have only drawn
two of the branches in each whorl. Even so the later
diagrams become very much complicated with all the
branching and rebranching. The spruce fir shows this
same regularity of shape, and so do several of the other
pines, hence you can tell their age in the same way.

The spruce fir (familiar to you as a Christmas tree)
is the evergreen that stands next in importance to the
Scotch pine. Although these two trees are rather similar
in shape when they are young, the spruce does not tend
to lose its pointed shape as it grows old. Its leaves are
much shorter than those of the Scotch pine and are
arranged singly along the twigs instead of in pairs.

The Pine family contains other members, as the
stone pine, the Weymouth pine, and the silver fir
(also one non-evergreen tree—the larch). Although
these trees differ from each other in details of form
or leaf, they all show the same general adaptation to
their environment of which I spoke at the beginning
of the lesson.

The timber of these trees is known as 'deal' and,
especially in the case of that of the Scotch pine, is very

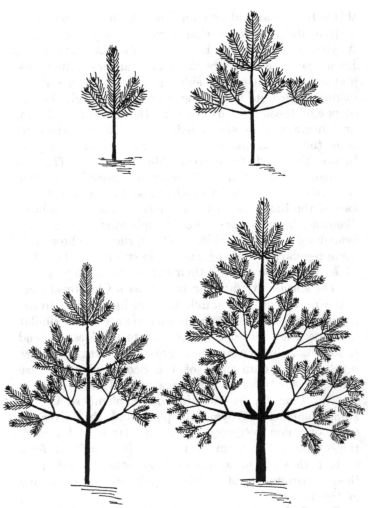

Fig. 35. Young Scotch Pines of two, three, four and five years old. Each year a new ring of branches is added to the main trunk and to each side branch. (In the last tree two of the lower branches have been cut off for the sake of clearness.)

highly valued, both for the good qualities of the wood itself and for the length and straightness of the tree trunks.   Deal is also the wood used in the manufacture of wood pulp from which many of the cheaper kinds of paper are now made.   If a slit is made in the trunk of a pine, turpentine will gradually ooze from it.   This is the commercial way of obtaining turpentine and large quantities of it are collected yearly from the Russian and American forests.

PRACTICAL WORK.

You will be taken for a walk to a pine wood.   While there notice the following points :—

Is there more than one kind of pine in the wood ?   Collect some pine needles from under each kind of tree.   Are they all single or are two or more of them joined together in some cases ?   Are there any seedling trees growing ?   If so, find out the age of some of the smaller trees.

When you get back to school you will be shown pieces of the timber of various conifers :—Scotch pine (red deal), spruce (white deal), larch, etc.   Notice how these woods differ in colour, grain, knots, etc.

HOME WORK.

Draw the twig you took home with you twice its natural size. Now take off a single leaf and draw both sides of it (also × 2).   Paint your drawings with green and brown colour washes and pick out the darker parts with deeper shades.   Write the name of the tree under your drawing.

CHAPTER VIII

HOW A TREE LIVES

LESSON 1

Season. Second week in February.

Materials required for each pupil (or each pair of pupils in the first and second questions).

Two Bunsen burners. Two tripods. One piece of wire gauze. One sand bath. Several four oz. flasks or beakers (about 3 in. deep). Porcelain dishes (about 3 in. across). Watch glasses. Pestle and mortar. One funnel and stand. Filter papers. Glass rods. Distilled water. A small heap (about a salt-spoon full) of salt, soda, borax, nitre and other substances to be examined, on separate pieces of paper with the name on each. Two glasses. One soft and one woody twig. Measuring glasses. One celandine or wild arum leaf. Pieces of narcissus stalk about 2 in. long.

If you were asked what things are most necessary to keep a man alive you would answer air, food and water, and you might add that for him to be healthy he also needs plenty of sunlight. The same thing is true of trees and of all other green plants. Part of the food that a tree needs is to be found in the earth, and, as it has to pass all over the tree, it can only be used when it is dissolved in water.

Put about a teaspoonful of salt into a glass of water and stir it well up. In a few minutes it has quite disappeared and the water is clear again. Taste the water and you will find that the salt is still there although it cannot be seen. If a little of this water is heated until it dries up, salt is left behind, almost unaltered by having been dissolved. The same thing can be done with nitre, alum and with several other things. Now try to do it with chalk or with fine sand. However long you may leave them in the water they only settle on the bottom of the glass, and when you stir the water up they are still there. In other words, substances like these will not dissolve in water.

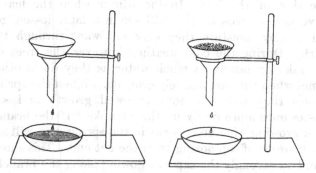

Fig. 36. A substance **soluble** in water cannot be separated from the water by filtering, but passes through the filter paper. An **insoluble** substance remains behind on the paper.

The water that is always to be found in large or small quantities in the ground contains dissolved matter of many kinds which the rain has washed out of the earth. When the tiny root hairs of which we spoke last term suck up water, they suck up with it these dissolved substances, and only those that are so dissolved; for no others will pass through the walls of the root hairs and into the root. The solution inside the root is clear and almost colourless and is called **sap**.

5—2

The sap, containing the food a tree needs, passes from the root hairs into pipes in the root. The pipes run all over the tree and take the sap up the trunk, along to the very tips of all the branches and into the leaves. In the leaves these pipes form a complete network of veins, so that no part of the leaf is left without its supply of food and water.

The leaves keep all the dissolved food and as much of the water as they need. The rest, in the form of invisible moisture, passes out through tiny openings in the skin of the leaf. In the winter when the leaves have fallen, trees, as you will see in a later lesson, get rid of any moisture they may not want through the bark. During the cold weather, however, the trees do not take up nearly as much water as they do at other times when they are actively growing. When the spring comes they wake up into renewed growth and sap passes once more freely up their trunks. If the branch of a growing vine tree is cut in the spring, you will see great drops of sap oozing from the cut end. This shows you how strongly the sap was flowing along the branch.

The sap, then, contains all the soil food which a growing plant needs. If there were no water in which this food could be dissolved a plant would in time starve, were it not for the fact that long before this time comes it would have withered away for want of water.

All the different parts of plants, stem, leaves, fruit, etc. are made up of countless tiny boxes placed side by side. The sap soaks through the walls of these boxes or **cells** and keeps them swollen and hence firm. If a young green plant is left many hours without water the walls of these cells collapse and the plant hangs its head and becomes flabby and withers. In a twig the walls are stiffened so that they stand firm however little water there is.

PRACTICAL WORK.

Place a young shoot and a twig in a glass without water. You will want them in question 5.

1. Test the solubility in water of washing soda, borax, nitre, etc. Proceed as follows : look closely at the substance and crush it up in the mortar with the pestle. Pour it into the beaker, and fill the latter about three-quarters full of hot distilled water. Place it on the wire gauze and let it boil for some minutes, stirring occasionally. Fold the filter paper in half and then again in half, and fit it into the funnel and moisten it. Fasten the funnel into the stand. Place the porcelain dish under it. Stir up the water in the beaker and pour it into the funnel. When it has all filtered into the dish, place the dish on the sand bath and let the water gradually boil away. While waiting for this you can go on with the next question. When the dish is dry examine it to see if the water has left anything behind. If so find out if it is the same as what you had at the beginning. According to whether there is left much, little, very little, or none of it write *soluble, slightly soluble, very slightly soluble* or *insoluble* next to the name of the substance on the paper. (Fig. 36.)

It is sometimes quicker to take just a few drops of the filtered liquid in a watch-glass and put the watch-glass either on a radiator or over the steam escaping from a flask of boiling water. The less water there is to evaporate the sooner you will find the answer and be able to go on to the next question.

2. Try the same test with garden soil, leaf mould, sand, peat, clay ; also with the substances used as fertilizers, crushed bones, kainite, mineral superphosphate, lime, sulphate of potash and magnesia; also with any chemicals employed in the botanical laboratory for water culture solutions such as potassium nitrate, calcium phosphate, magnesium sulphate.

3. Prepare a series of salt solutions of various strengths. First a solution containing 32 grms. per litre. This will taste about as strong as sea water. Take 500 c.c. of it and dilute with 500 c.c.

of water, making a solution of 16 grms. per litre. Dilute half of this until you have a set of solutions of strengths 32, 16, 8, 4, 2 and 1 grm. per litre. Obtain some narcissi or other juicy stalked flowers from the shops. Slit up the stalks of 7 flowers from below with slits about 1 or 2 inches long. Put one flower stalk in each solution and one more in distilled water. Leave them for a day, and describe what happens in each case.

4. Place the blade of a celandine or wild arum leaf in water. Suck hard at the cut end of the stalk. What change do you notice in the colour of the blade ? How do you account for it ?

5. What has happened to the soft and the hard stem you left in a glass without water ? Fill the glass with water. What do you expect to happen now ?

6. Use white flowered narcissus if obtainable in the shops. Put their stalks in red ink, and leave them for a day. What happens ? What do you see when the stalks are cut across ? Try black ink and coloured inks as well. Try other flowers ; and try twigs freshly cut from living trees.

HOME WORK.

Place a young twig about 9 in. long in water coloured with red ink, having first removed a ring of bark ½ in. deep, about 2 in. from its lower end. Leave the twig in water until the next evening. Then take it out of the water. Sketch it life size. Now, beginning at the upper end cut off ½ in. at a time to find out how far up the red ink has reached. When you come to the red ink notice what part of the stem it passes up. Has the removal of the ring of bark stopped it ? Sketch the piece you have left by the side of the whole twig.

LESSON 2

Season. Second week in February, but revise experiment 5 in June.

Materials required. (*a*) **For Demonstration.** *Chemical lecture room and laboratory useful but not essential. Small saucer or crucible. Basin containing water coloured with red ink. Glass jar* (*e.g. a jam pot*) *which will go comfortably into the basin. Small wad of cotton wool. Methylated spirit. Lime water. Small flask with tightly fitting rubber cork. Vaseline. Glass tube* (U-*shaped as in Fig.* 37) *to fit tightly into cork. Test-tube which will go into the flask, containing* ½ *doz.*—1 *doz. peas or beans which have been soaked in water for* 12 *hours.*

(*b*) **For each pair of pupils.** *Small flask half filled with lime water. Cork fitted with two glass tubes, longer of which passes into the lime water at one end and is fitted at the other with* 9 *in. of rubber tubing. Pair of bellows. Two fresh leaves, part of whose blades have been covered with tinfoil for* 24 *hours before picking. Bunsen burner. Tripod stand and wire gauze. Glass rod. Beaker of water. Two small glass dishes, one containing alcohol and the other alcoholic solution of iodine diluted with water to a pale brown colour. Dish of plain water.*

To understand how and why a plant breathes you must first know something about the air with which we are surrounded.

This air is of three sorts, all mixed together :—the first gas, **oxygen**, is necessary to us for life and is very strengthening. It would however be too strong for us if we always breathed it pure. **Nitrogen**, the second, is not able to support life and serves only to weaken the effect of the oxygen. Besides these two gases even the purest air contains small quantities of a poisonous substance called **carbon dioxide** or **carbonic acid gas**. When people, animals, and, to a lesser extent, plants breathe they take in oxygen and nitrogen from the air, use up the oxygen, and breathe out the nitrogen together with a good deal of this poisonous carbonic acid gas. In this same way all burning things use up oxygen and make carbonic acid gas and moisture. This is the reason why if you are for some time in a room with a lot of people or gas jets the air soon becomes stuffy and you will get a headache if you do not open a window and let in some fresh air.

Now as for so many years all the people, animals and plants in the world, to say nothing of the fires, have gone on making this poisonous gas and using up the oxygen you might think that the air would have long ago become unfit to breathe. This is prevented by the *leaves* of trees and other green plants. All day long they are busy making the air wholesome by taking from it the carbonic acid gas which is to them a most useful food.

This carbonic acid gas is made up of two parts, carbon and oxygen, bound together so tightly that it is almost impossible to separate them. Green leaves have, however, wonderful chemical powers. Helped by light which is as necessary to them as heat is to a steam engine, they split up this gas, take the carbon which is what they need to build up the plant, and send the oxygen that has been set free out into the air. You see how enormously valuable they are to us, for they are in this way not only taking the poisonous part from the air, but are actually adding to it more of the wholesome part that we want to breathe, the oxygen. I have said that this useful work can only go on in the light. In the dark it stops and the other process, that of using up oxygen and breathing out carbonic acid gas, which has been going on slowly all day, goes on alone. This is the reason why it is not a wholesome plan to leave plants in a sick person's room during the night, though during the day they not only brighten the room but help to keep the air pure. Water plants growing at the bottom of streams get the carbonic acid gas they want out of the water they grow in as there is always some dissolved in it. If you watch a water plant in an aquarium in the bright sunshine you will see a stream of little bubbles coming up through the water. This is the oxygen which the plant is setting free.

You will want to know what becomes of the **carbon** which trees are always collecting. Some of it is used to make the wood of which the tree is built up. The black substance which is left behind when you burn a twig or a match is carbon or charcoal and shows you what a lot there is in wood. Some more of the carbon is turned by the leaves into **starch** and is stored up in the leaves or roots to be used as food in the winter and spring when the tree needs it. By using a certain stain which turns starch blue you can see for yourself how a leaf that has been in the sun is full of starch, while one that has been in the dark for some hours has had to use up all its starch as food and has not been able to make any more because, as I have told you, the leaf cannot do its work of collecting carbon in the dark.

PRACTICAL WORK.

1. (**Demonstration.**) Some peas which have been soaked for 24 hours are placed in a test-tube containing just enough water to keep them moist. The test-tube is placed in a flask containing a little lime water and fitted with a rubber cork. One end of a glass tube passes through the cork into the flask, while the other forms a U-shaped curve and is filled with coloured water. The cork is placed in the flask and the whole is made air-tight with vaseline.

As the peas grow they will use up some of the oxygen in the flask
(Fig. 37). The carbonic acid gas which they breathe out is absorbed
by the lime water, hence the total air in the flask grows less. The
original level of the water in the tube must be marked with a label.

Make a sketch of the apparatus used and mark in it the original
level of the coloured water. Look at the flask again to-morrow and
see if there is any change in the level of the water. If so mark it
in your sketch.

Is there any change in the lime water ?

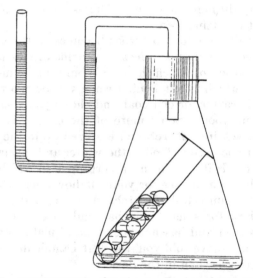

Fig. 37. **Practical work.** Question 1. To show that germinating
seeds use up part of the air.

2. Moisten the indiarubber tubing leading from the flask of lime
water, and place it over the nozzle of your bellows. Blow air through
the bellows into the flask. Do you notice any change in the lime
water ? Now take out the bellows and place your mouth to the
tubing. Blow air from your lungs into the lime water. Do you
notice any change now ? N.B. The air passing through the bellows

is pure while that coming from your lungs is charged with carbonic acid gas.

3. (**Demonstration.**) Some cotton wool is soaked with methylated spirit and placed on a small saucer floating on coloured water. The wadding is set alight and is quickly covered with a glass jar. Notice what happens as the wadding burns. The air that is left in the glass jar is shaken up with a little clear lime water. What happens to the lime water ? Write an account of the experiment in your own words and explain what you learn from it, giving diagrams.

4. You are given leaves, parts of which have been kept from the light for 24 hours by coverings of tinfoil. Boil them in water, but only for one minute, this kills them ; then place them in alcohol to get rid of the green colour. Now if you put them into iodine you will find that those parts of the leaves which were exposed to the light turn dark blue, showing the presence of starch. Put your stained leaves into clear water and leave them for inspection. (This experiment with leaves partly covered with tinfoil should be repeated in the summer when the sunlight is brighter and summer leaves, *e.g.* clover or nasturtium, can be obtained.)

HOME WORK.

Write a short essay, not more than three pages, on " How Plants feed and breathe."

CHAPTER IX

THE ROOT

Season. Third week in February.

Materials required for each pupil.

A mustard seedling seven days old. (The seedlings can be grown in moist air by fitting a piece of moist blotting paper inside a tumbler, and pushing the seeds down about 1 inch from the top of the tumbler and between it and the blotting paper. The paper is kept moist by pouring a little water into the bottom of the tumbler.) A geranium cutting taken in October. (These cuttings should be about 3 in. long when taken, and should be cut sharply through just below a joint. They must be placed in pots or well drained boxes containing sandy garden soil (about 2 parts soil and 1 sand) and must be sheltered from the frost during the winter months. They want little or no watering.) The cuttings must be well washed to free their roots from soil. If they have not been specially arranged for, they can probably be bought from a florist. A broad bean placed a fortnight ago in a test-tube of water, as directed in Chapter IV. A small drawing board. Sheet of drawing paper. Drawing pins. 2 yards of clothes line rope. Tumbler (or test-tube) of water. Cork. Pins.

Now that we have spoken of the processes by which a tree lives, we will consider the parts of the tree in turn and see how Nature has suited them for the work they have to do, and how they vary in different trees so as to accommodate themselves to the conditions under which the tree lives.

A tree's **root** has two important parts to play :—In the first place, as we saw in the last chapter, it supplies the tree with the water so necessary to its existence and with minerals dissolved in it. Our seedling oak tree showed us how smaller roots grow from the sides of the main root and then still smaller ones from these. In a young tree the rootlets grow quite regularly to begin with, but, later on, outside influences interfere in the shape of slugs, frosts and irregularities in the soil and so the root grows gnarled and crooked.

While the trunk of a tree increases in size and in thickness the root goes on growing underground until it reaches for yards and has formed quite a network of smaller roots reaching in all directions. The principal root goes straight down into the ground. The roots next to it in size grow out from it pointing slightly downwards, and they in turn give out smaller branches, so that among them they thoroughly explore the whole of the ground covered, and manage to extract water from earth which, to look at, anyone would say was absolutely dry. Of course the earth is often only sun-dried on the top and the deeper roots grow the more water they will find. Thus the oak which needs a good deal of water to grow well makes sure of it by sending roots right down into the earth, while trees which, like the firs do not want much, have roots which branch and spread

Fig. 38. Young Oak. Dicotyledons have a principal **tap root** from which the other roots branch.

Fig. 39. Wheat seedling. Monocotyledons have a number of roots of equal size and importance instead of a tap root.

out near the surface. It is because of these shallow roots that the fir is able to grow among rocks on the mountain side where there is not much depth of soil to be found.

In planting a wood of mixed trees it is well to remember that some have deep roots and others shallow ones, and that as the two kinds exhaust different layers of soil they can be planted together more closely than they could be if this were not the case. Again if trees are to be planted along the edges of a field, deep rooted trees like the oak and elm are better than the often seen poplar, as the latter's roots spread out near the top of the field and hence use up the very soil that will be needed by the crops. The extreme example of these shallow roots are those of the ash which grow so close to the surface that hardly anything will grow under an ash tree. People used to think that this was because the ash poisoned the ground, but the only reason really is that its roots have used up all the goodness in the upper layer of soil so that other plants starve.

The second way in which the

root serves a tree is by anchoring it firmly to the ground
—no simple task when you remember the masses of
leaves borne by some trees and the violent gales which,
in spite of these, they successfully resist. A tree which
like the birch or poplar has a flexible trunk easily swayed
by the wind does not need such
strong deep growing roots as the
oak which does not yield to any
gale until actually uprooted from
the ground.

It has been suggested that
builders first took the idea of the
buttresses with which the walls
of churches and cathedrals are
strengthened from the way in
which a tree's roots grow out from
the trunk. Again, if you compare
these roots with the guide ropes of
a tent you will see how well they
are adapted to bracing up the tree.
If they grew out in one direction
only any strong wind could blow
the tree over sideways; they
branch however on every side
equally. From these two instances
you will see that this second use
of the root is altogether mechan-
ical, that is to say it does not
depend on the root being alive,
but only on the way in which it
reaches down into the ground.

Fig. 40. Wheat seedling,
with soil sticking to the
root hairs.
N.B. *The root hairs do
not grow right to the
ends of the roots.*

Besides supporting and nourishing the tree itself,
the roots serve the additional purpose of binding the

soil firmly together. In countries where shifting sand
dunes make cultivation of the land impossible one of
the first things that is done is to plant quantities of one
of the pines (*e.g.* in the Landes, a sandy district in the
South-West of France, the cluster pine is used). These
trees grow quite well on dry sandy soil and their thickly
branching roots soon hold the sand together and prevent
it from shifting, while the pine needles cover the ground
and gradually improve the soil. Trees are also useful
where the land is subject to floods, as the roots bind
the loose soil together and enable it to resist a sudden
rush of water.

PRACTICAL WORK.

1. Draw your mustard seedling × 4, paying special attention to
the root hairs. If these are stuck together you can pin the seedling
to a cork and float it in water.

2. Examine the root of your bean plant. Do the rootlets seem
to grow out from the root in any definite way or not? Give a
drawing of the root to illustrate your answer.

3. Explain what has happened to your geranium cutting in the
past three months. What do you think would have happened if you
had put it into the ground the other way up?

4. Draw a map of the school garden, marking the position and
name of each tree. Tie your piece of rope into a loop, sling it round
your neck, and use it to support the drawing board. With a little
adjustment of the knot the drawing board will lie level and you can
use it as a portable table to draw on as you walk about. The map
you make now may not be perfectly accurate, but you can be careful
to make it neat and to get the positions of the different trees as
nearly right as possible. [For details about map making see books
on Practical Geography.]

HOME WORK.

Make a fair copy in ink of the map you made in the garden.
Bring both copies to school with you.

CHAPTER X

THE TRUNK

Season. Fourth week in February.

Materials required for each pupil.

A section of wood showing rings, with bark attached (see Appendix). A mounted wood shaving (see Appendix, or those from a carpenter's workshop may be used). A narcissus head with long stalk. Glass tube. Twigs of beech, elder, Scotch pine, horse-chestnut, willow, oak, to show lenticels.

For Demonstration. *Beetroot. Natural uncut cork. Oak bark as before use in tanneries.*

If you look at the stump of a tree which has been cut down you will see a number of rings one inside the other. These rings consist of bundles of pipes up which the sap passes on its way to the branches, and of strengthening fibres, and their number depends upon the age of the tree. Each year it adds a new ring outside the old ones, and the sap passes up the new ring as well as up the others. After a certain number of years the rings in the middle of the trunk dry up. The old wood, although no longer alive, is kept from decay by the living ring round it and acts as a girder to hold the trunk together and help it to bear the branches and masses of leaves. It is called the **heart wood**, while the new ring through which the sap flows is the **sap wood** (Fig. 41). The heart wood is darker in colour than the other, and, as it is harder and more durable, is more valued. A thin shaving of wood will show you very nicely the old pipes and fibres lying close together, all growing in the same direction. This direction is what carpenters mean when they speak of the **grain** of wood. When they use a plane for smoothing planks they work it along these fibres and never across. This is also the only direction in which wood splits.

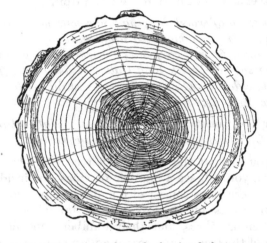

Fig. 41. Cross section of an Oak trunk, showing dark **heart wood** in the centre with lighter **sap wood** surrounding it.

As a ring is added to the trunk every year, the older a tree is the thicker it will be, and you can find out its age by counting the number of rings. Not only this, but, by looking at each ring you can tell by its width if the weather during the year when it was added was mild or severe. A warm moist year will produce a thick ring, and a cold or dry year a thin one. If the roots at any time happen to reach an unusually rich soil, it will have the same effect as a warm season and will produce a thick ring. The rings you see on a slice of beetroot grow in just the same way except that a new one forms every fortnight instead of once a year.

What is known as **bark** is really the dead outer skin of the tree. This outside layer cannot stretch to fit the growing size of the trunk and soon becomes too small. It therefore cracks or in some cases peels off in scales, and new bark is formed underneath to take its place. The thick skin which the old and new bark form together is of use to the tree in two ways :—it prevents the tree from losing too much water through the trunk in dry weather ; and it guards from the direct heat of the sun and from frost the delicate growing part of the tree which lies all round the trunk just under the bark. The younger branches and twigs sometimes suffer for want of a thicker layer of bark than they have, for, although during the summer the leaves take its place and serve to shade them from the sun's rays, when these leaves are gone, frosts which do not hurt the large branches under their thick coat of bark often nip the twigs.

Bark depends very much for its thickness on the requirements of the tree on which it grows. Since one of its uses is to protect the trunk from the sun, when any kind of tree usually grows in the sun it will have a thicker bark than the shade-seeking ones. At first sight you might think that the papery looking bark of the birch, the most sun-loving tree of all, was an exception to this rule (Fig. 56). This bark, however, consists of many thin layers one beneath the other and forms a better shelter than one thick coat would do. Besides this some people think that its white colour helps to keep the tree cool.

The best example of a thin barked tree growing in the shade is the beech whose smooth thin bark you all

Fig. 42. Trunk of Plane tree with scaly bark.

know. If by any chance one of these trees has grown all its life in a very sunny position it is found that its bark tends to become thicker than it usually is. On the other hand if trees which, like the oak, generally have a thick bark, are grown among a lot of others the bark is apt to become thin and soft. This principle has to be remembered when plantations are being thinned, as, if this is done too suddenly, trees which have till

Fig. 43. Trunk of Oak tree with fissured bark.

then been sheltered by others growing round them will suffer from the sudden exposure to the sun.

There is one form of bark which is of great commercial value. **Cork** is the inner bark of the cork oak which in this tree grows enormously thick. The many everyday uses to which it is put depend partly on its lightness and partly on the fact, so important to the tree, that water will not pass through it. All inner bark is corklike in nature although it is only in this particular tree that it is found in sufficient quantity to be of practical use to us. It is full of little air holes which supply the tree with air in the winter when the leaves by which it usually breathes are gone. The openings to the outside air are called **lenticels** and are quite easy to see on most twigs.

Cork is the covering with which a tree always protects itself after any injury. If one of its branches is cut off close to the main stem the inner bark soon begins to spread itself from the sides over the wound, and if this wound is not very large the cork may succeed in covering it altogether and the tree will be none the worse. Otherwise rain will get in and may very likely begin to rot away the wood. Then in time the tree will become hollow until at last only the bark and a thin ring of sap wood is left. You see what serious harm it may do a tree if its bark is injured. Unfortunately some animals, especially deer and rabbits, are fond of gnawing the bark of young trees, especially when the ground is covered with snow. This is one of the dangers from which newly planted trees in plantations have to be carefully guarded, or the bark will be injured and the tree will probably die.

PRACTICAL WORK.

1. Sketch two views of your log, one showing a transverse section of the wood and one a longitudinal one.

2. How old was the tree from which your log was taken ? Had it lived through any extremes of heat or cold ? How do you know this ?

3. How many different kinds of layers can you distinguish on your log, counting inwards from the outside bark ?

4. Pass a glass tube into the stalk of the narcissus. Fasten it firmly with wool. Put the stalk, flower end downwards, into a glass of water. Blow gently into the tube. What happens ?

5. Examine carefully the different twigs. Write down the name of each twig. Under the name write whether lenticels are present or not, and if so whether they are large or small.

6. Draw your mounted shaving. Notice whether the space between the different layers is equal, or nearly so. Indicate in your sketch by an arrow the line in which you would use a plane.

7. Write down uses to which at least three different parts of the oak tree are put.

HOME WORK.

Make a water colour sketch not less than 9 in. high and 2 in. wide of the trunk of a tree, showing where the trunk meets the ground and where the branches begin. Write the name of the tree underneath.

CHAPTER XI

CATKINS

Season. About second week in March. (This lesson must be interchanged with one of those preceding or following it if the year is a forward or late one.)

Materials required for each pupil.

Lens. Dissecting needles. One twig showing cat-kins of each of the following trees (or two if male and female flowers grow on separate trees): sallow, alder, poplar, bog myrtle, walnut, beech, oak, Spanish chestnut. (The last three will not be in flower as early as the others, but may be used for revision in a few weeks' time.)

In the early spring, before the leaves are out, some of the trees are already breaking into flower. If you look in the hedges at about this time of year you will find that the bees are clustering on those silver and golden tree flowers that country children call "palms," and the bare twigs of the hazel bushes where you gathered nuts in the autumn are showing dangling **catkins**, as these long hanging bunches of flowers are called (Fig. 44). We will look at some of these bunches more closely and try to find out what a catkin really is.

You will see that the hazel catkin is made up of a number of tiny flowers growing round one long stem. Each flower is too small for you to be able to see it distinctly with your eyes, unhelped by any glass, but you can take a few of them carefully off with the point of a long pin and look at them through a pocket lens. Each of them is made up of a greenish scale and of a bunch of **stamens** covered with a golden powder called **pollen** that shakes off when the catkin swings in the wind (Fig. 44 *a*). Now look carefully at the rest of the twig. Do you see anything on it besides the long stamen-bearing catkins? Here and there is what looks like a green bud with a few crimson hairs sticking out at the top. This is a bunch of very simple flowers of another kind. Break off one of these buds and take it very carefully to pieces. The scales of which it is made up come apart and some of the inner ones are found to have one forked red thread fastened to each. This thread ends in a swelling at the lower end and is called

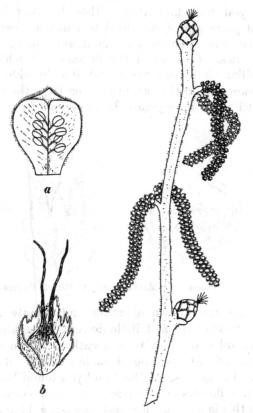

Fig. 44. Hazel twig and catkins.
(a) Male or *stamen* flower.   (b) Female or *pistil* flower.

the **pistil** (Fig. 44 b). If some of the pollen from the long catkins falls, or is blown by the wind, on to the threads of the pistil, each of these pistil flowers will in due course bear a hazel nut.

Next term when you learn about other kinds of flowers you will find that, although other brightly coloured parts are often added to make a flower more attractive, the stamens and pistil are the really important ones. In most of the flowers with which you are familiar these two are found side by side in the same flower, but in the catkin-bearing trees the stamens and pistil grow in separate bundles and form respec-

Fig. 45. Hornbeam. (*a*) Male flower. (*b*) Female flowers (two).

tively what are known as **male** and **female** flowers. When a great number of little flowers of one kind grow on a central stalk they form a catkin, and so you get two kinds of catkins, one of male and one of female flowers. In the case of the hazel you found both male and female flowers on one tree, but in some other kinds of trees this is not so. For instance the gold and silver catkins of the **sallow** or **willow** grow on separate trees and the pollen is carried from the male to the female flowers by the bees which gather honey first on one tree then on another. When their pollen is gone the work of the male flowers is over, and the whole catkin gradually dries up, and then falls to the ground. The female catkins last until the seed has ripened.

Besides the male or female flowers you will often find on a catkin a number of small scale-like leaves in the axils of which grow the flowers, exactly as a leaf bud grows in the axil of a leaf. Leaves that bear this relation to flowers are known as **bracts**, and their presence on a catkin sometimes makes it much harder to see where one flower leaves off and the next one

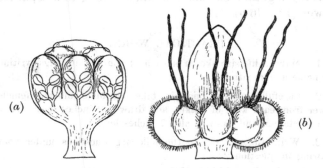

Fig. 46. Birch. (*a*) Male flower. (*b*) Female flowers (three).

begins. To show you what I mean we will look together at the catkins of the **hornbeam**. In this tree the male and female catkins are found together on the same twigs. Each flower of the former includes a greenish bract, somewhat curved over the bunch of forked stamens which grow upon it (Fig. 45 *a*). The female catkins bear a number of long green bracts under each of which are *two* flowers. When these are separated they are not unlike the hazel flowers. Each one consists of a pistil with two long red ends growing from a hairy bract (Fig. 45 *b*). In the autumn pistil and bract will together form a winged fruit.

The flowers of the **birch** are yet harder to understand. The male ones are made up of no less than six bracts. First a big one behind, then two together, and three in front of these. The twelve stamens grow on the three front ones (Fig. 46 *a*). The female flowers are arranged in groups of *three*, all growing from one bract. We know that this is so because one flower cannot have more than one pistil, and here are three pistils growing together, so each one must belong to a separate flower (Fig. 46 *b*).

PRACTICAL WORK.

1. Make a life-size sketch of each of your catkin twigs, writing the name underneath.

2. Carefully, with a long pin, take one male and one female flower from each kind of catkin, and draw it 2 inches long. Take off one stamen and draw it also 2 inches long.

3. Write the answers to the following questions under your drawing in question 2:

    (i)   Do the two kinds of flowers grow on the same twig or on separate ones?

    (ii)   How many bracts are there to each flower?

    (iii)   How many stamens are there in the male flower?

    (iv)   Is the flower scented? and can you find any trace of honey on it?

HOME WORK.

Take home your twigs and the sketches you made of them, and carefully paint the sketches in water colour.

CHAPTER XII

THORNY AND CLIMBING PLANTS. (A WALK)

Season. About third week in March.

Materials required.

Twigs of various prickly, spiny and thorny plants and climbers other than those mentioned in the lesson will be wanted for home work. Each child should collect them during the walk and take them home.

Most of the small four-legged inhabitants of our woods and fields are vegetarian in their tastes, hence any natural protection which will keep them away from a plant will be of great service to it. This protection takes several forms. Sometimes the leaves are unpleasantly bitter or even poisonous, in other cases the plant is provided with bristly hairs or with spines. How well these spines can serve their purpose is seen in the common **gorse** which, as it grows old, becomes such a mass of spines that the cattle, which liked to browse on the young shoots before the spines hardened, leave the older ones severely alone. During our walk to-day I want you to see some of the forms these thorny defences take.

One of the first examples that you will come across is that of a weed which grows everywhere and from which you have all learnt to keep away unless your hands are well protected. I am speaking of the **stinging nettle.** Its leaves and stem are covered with stiff hairs, and each of these hairs is hollow and contains a little

poison. If you seize a leaf firmly between your finger
and thumb you crush the hairs and they will not hurt
you. If, however, you let them touch your hand gently
they will pierce the outer skin and will leave a tiny
drop of poison behind. Then your hand smarts and
you say you have been stung.

Another leaf that you have probably learnt to handle
carefully is that of the **holly** tree. The surface of these
leaves is smooth enough, but a horny thread runs along
their edge and forms a series of sharp spines which
would soon send away any marauding deer or other
animal in search of a meal.

Perhaps the commonest type of prickly plant is that
of the **wild rose** and **bramble**. Look at a twig of wild
rose and see how plentifully it is furnished with thorns
—curved thorns as sharp and hooked as a cat's claws,
which when they get caught in anything are not easily
got rid of. Cut one of the twigs off close to the ground,
and try to pull it out by its lower end. You cannot do
it, because the thorns catch in the other branches. If
you take hold of the other end and pull it upwards it
will come easily enough. Notice that the thorns do
not seem to be arranged in any special order, but are
scattered about quite irregularly on the twig. Try to
pick off one of the larger ones, using your fingers or
a penknife if you have one. It comes off quite easily,
leaving a mark on the outside of the stem. Take
several of these thorns off, then peel the skin from the
twig. There is no mark at all on the wood of the stem
to show where they grew. This tells us that they were
only thickenings in the skin, and were not part of the
stem itself.

Now take a twig from the next **hawthorn** bush you

see and look at its thorns (Fig. 47). They are straight, very sharp, and some of them are much longer than those of the rose. You will find that you cannot separate them from the main stem without cutting through the wood. Moreover, if you peel the twig, you will see that the skin covering comes off and leaves

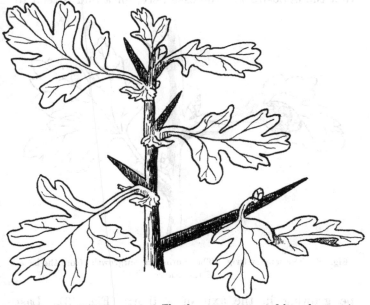

Fig. 47. Hawthorn twig. The thorns are converted branches.

behind woody thorns which, though smaller than they were before, are still the same shape. This time, therefore, the thorns are not merely on the skin as in the case of the rose. What are they? Do you see that the longer ones have leaf buds growing from them? This shows us that they must really be branches which have become sharp at the end to serve as a defence to the trees. Another proof that this is so lies in the fact that the thorns are seen to grow in the axils of leaves, and you learnt last term that this is where the young branches are always to be found.

On looking at a **gooseberry** twig you might think that the opposite was the case here, for a bud is seen to

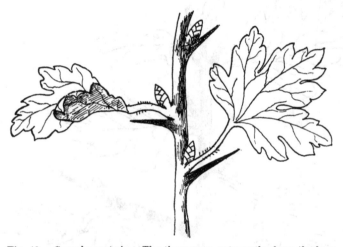

Fig. 48.   Gooseberry twig.   The thorns are outgrowths from the base of the leaves.

be growing in the axil of a thorn (Fig. 48). Last autumn, however, you would have found that a leaf grew between the thorn and the bud, and if you look carefully now between the two you may be able to see the leaf scar there. The thorn cannot therefore be a changed leaf, since the leaf was there as well. All you can say of it is that it is a woody outgrowth from the base of the leaf.

When you looked at the thorns on the wild rose
you noticed how their claw-like shape made their under
sides catch in the other branches as soon as they touched
them. Does this suggest to you a second important part
that is played by thorns like these? In the thick
undergrowth among which briars and brambles usually
grow each plant has a hard struggle to get as much
light and air as it wants, and anything that will help it

Fig. 49. Brambles growing among their natural surroundings in a wood.

to raise its head above its neighbours will give it a great
advantage over them. This is just what the thorns do
for the briars. As the young shoots grow taller the
thorns on them seize firm hold of any hedge or bush
against which they may be growing and prevent them
from slipping back. Other plants have various ways of
their own of gaining the same end. The long flexible
stem of the **ivy** is very dependent on outside support,

and when this support is found in the shape of a convenient tree trunk little roots begin to grow all up one side of this stem. These roots grow away from the light, as all roots do (see Chapter V), hence they grow towards the tree trunk, of which they take firm hold. The ivy like any other plant draws all the food it wants from the earth and air, so that it draws no nourishment from the tree up which it grows. Nevertheless if it is allowed to go on growing unchecked it will soon cover the trunk and will seriously weaken the tree by keeping light and air from it. A good landowner is therefore careful that his trees are occasionally freed from all ivy.

A large number of climbing plants (including under certain circumstances the ivy itself) have twining stems which grow round and round any support they meet, and so hold themselves upright. In other cases the leaf stalk is twisted like a corkscrew and the plant climbs by its help. To the first of these classes belong scarlet runner beans and hops, to the second the clematis. Other climbers again, as the bryony and the various vetches, make use of long curly organs called **tendrils** which are really different parts of the leaves or stem converted specially for this purpose.

PRACTICAL WORK.

You will be taken for a walk and must notice in the woods and hedgerows the various plants mentioned in this lesson. Notice the conditions under which they grow. Collect any thorny, spiny, prickly, or climbing plants that have not been used in your lesson (*e.g.* mahonia, barberry, robinia, blackthorn, convolvulus) and take them home.

HOME WORK.

Make a life-size sketch of each of the twigs you brought home. If a sketch represents a thorny twig, write under it what you think the thorns really are, giving your reasons. If it is of a climber, state the part used for climbing.

CHAPTER XIII

LEAF BUDS

Season. Fourth week in March.

Materials required for each pupil.

One Brussels sprout. One horse-chestnut bud. One beech bud. 4—6 living twigs of various kinds and ages. Pencil and writing block for writing in the garden. Dissecting needles (large blanket pins or a small penknife can be substituted for these).

There is one form of leaf bud that you all know very well. Pull the curled leaves of a Brussels sprout off, one by one. They grow smaller and smaller until they are almost too small to see at all. At last they are all gone and a short fleshy stalk is left behind, with knots that are really tiny new buds growing in the axils of the leaves you pulled off. The whole sprout is merely a **bud** with young leaves folded over each other to resist the cold weather.

Buds vary very much in colour and in shape in different kinds of trees. If you look at them during the winter you will quite soon get to know the round black ash buds and the pointed reddish ones of the

beech tree. Others that are easy to recognise are:—
oak buds, small, bright brown, and very crowded at the
end of the twig—sycamore buds, green in colour and

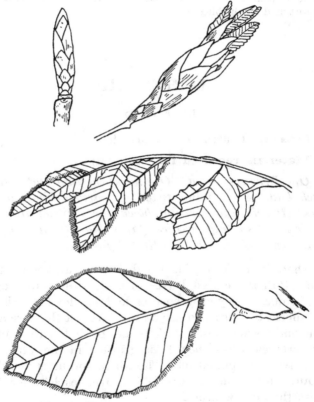

Fig. 50. Four stages of an opening Beech bud.

rather square and solid looking—**horse-chestnut** buds,
which are sticky and large, those at the end of the
twigs being especially big. We will take one of these

last and see how beautifully it is adapted to its purpose of sheltering the delicate young leaves and flowers. The sticky substance with which it is covered protects it from rain and holds the scales which form the outer layers firmly together. Pull off these scales carefully with a pin and notice that they are arranged in pairs, each pair overlapping the pair beneath it. The outer scales are brown. Then they become green and rather woolly. When you have taken off about six pairs you come upon two little leaves, each with five fingers and covered with white fluff to keep them warm. There are several pairs of leaves, each smaller than the last, then, if your bud is big enough you will see in the very middle a tiny thing like a bunch of grapes which will in time be the horse-chestnut flower. We took one of the large end buds to see where the flowers grow, but the smaller buds, growing along the sides of the twig, are made up of leaves only.

Now take a **beech** bud and see how different it is in colour and in shape. It is much smaller too. If you take the scales off this one you will find first several layers of brown scales, then these scales gradually become green at the bottom and almost transparent and between each pair of scales there is a tiny leaf fringed with silvery down. If you are very careful you can take all the scales off without hurting the leaves, and a delicate bunch of leaves is left behind. Can you see how curiously folded the young leaves are when they are tucked into the bud?

On most trees the leaves and flowers are found in separate buds. This can be seen very nicely at this time of year on any apple or **pear** tree (Fig. 51). Some buds are pointed while others, the flower buds, are round and much larger. It is important for gardeners when they are pruning the trees to know the difference between the two, or they would cut away the flower

Fig. 51. Twig of Pear tree showing leaf and flower buds.

buds which would in time turn into fruit, and keep only the leaves.

Now let us see what will happen to our beech bud in a few weeks' time. If you stand a twig in water which

you are careful to change every two or three days, and
keep it on the chimney piece
or in some warm place, it will
go on growing and you will be
able to watch it. The leaves
will become larger and will
gradually show beyond the tips
of the scales, then they will
open out and all the scales will
fall off. During the summer
the spaces between the leaves,
called **internodes** (**nodes** are
the places where the leaves
grow), will grow longer and
longer until the bud has become a twig with leaves arranged along it. In the axils
of these leaves buds will form
in their turn and will next year
carry on the growth of the tree.
The twig which was last year
a bud still shows the marks
where the outer layers of scales
used to grow in the form of a
number of rings round its base
(Fig. 52). As these marks do
not wear off for a long time,
and as new buds are formed
every year, you can always tell
the age of a larger twig by the
number of sets of ring scars on
it, the space between any two
sets being one year's growth.

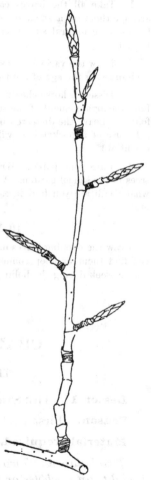

Fig. 52. Beech twig four
years old.

PRACTICAL WORK.

1.   Take all the leaves carefully off your Brussels sprout, and arrange them on a sheet of paper. Draw the 1st, 10th, 20th, 30th leaves twice natural size, also the axis when all the leaves are removed.

2.   Draw the various twigs natural size and write by the side of each drawing the age of the twig.

3.   Dissect a horse-chestnut bud and a beech bud, noticing carefully the arrangement of the scales and the way the young leaves are folded.   Leave the dissected buds on a sheet of paper for inspection. (The use of a pocket lens will be found to be helpful, but is not essential.)

4.   Date your paper. Write in column a complete list of the trees in the school garden. Write opposite the name of each tree in what condition you find it, *e.g.* 'leaf buds bursting,' 'in full flower,' etc.

HOME WORK.

Draw the opening leaves of any one tree in as many stages as you can find them, or keep some willow twigs in water and draw them once a week during the holidays.

CHAPTER XIV

THE LEAF

Lesson 1.   THE STRUCTURE AND PARTS OF A LEAF.

Season.   First week in May.

Materials required.   (*a*) **For demonstration.**

Two balances.   Two specimen glasses.   Cotton wool. Weights and pebbles or shot.   Two living twigs (a kind with large leaves gives the best results).

(*b*) **For each child.**

Leaf bud of common or wild cherry, or horse-chestnut just bursting into leaf. One of each of the following leaves:—apple or pear, honeysuckle, hawthorn, beech, elm, rose or blackberry. (In each case where an alternative is given the first named should be used if possible.) Twigs of lilac and laurel (or ivy) bearing leaves. Botanical pressing paper or blotting paper.

Do you remember that when we were speaking of the way in which trees and all other plants live you learnt that the green leaves are really a tree's lungs by which it breathes and gets rid of unnecessary moisture? I told you also that with the help of the sunlight they manufacture food for the tree in the form of starch. Since both light and air are needed for these processes a leaf has a broad flat blade which exposes to them as large a surface as possible. This blade is so thin that it would be very easily torn if it were not for the framework of veins by which it is strengthened just as an umbrella is by its ribs. The soft part of a leaf which lies in between the veins is made up of tiny cells packed together, each of which contains a number of round green bodies which give the leaf its colour and are responsible for making the starch. If a plant is grown in the dark its leaves are sickly and white because without light these green bodies cannot exist, but in a healthy plant they are always there. One case may occur to you which looks like an exception. How about red leaves like those of the copper beech? Even in these the green is really there, but it is hidden by the red colour of the outer skin. In some trees which like the *variegated* kinds of ivy and holly are grown for ornament the green bodies are arranged in patches, and then the leaves are flecked green and white. A red colour is often to be seen in young leaves and disappears as they become full grown—in their case this red skin serves to protect the delicate young from the sun's rays until it has grown strong enough to bear them.

The outer skin is harder and tougher than the inside of a leaf. This prevents the sap inside from being

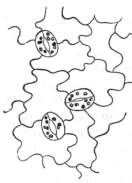

Fig. 53. Three **stomata** from the lower surface of a leaf.

dried up, as you will realize if you strip a piece of this skin from a leaf and notice how soon the leaf begins to fade without it. Dotted over the skin, especially on the under side of leaves where they are less likely to get choked by rain or dust, are thousands of little pores through which water vapour passes on its way from the inside of the leaf to the drier air outside. These openings are called **stomata**, from a Greek word meaning mouths, and are so tiny that they can only be seen through a microscope. If you are lucky enough to have one and look through it at the well scraped outer skin from the under side of an ordinary cabbage leaf you will see something like Fig. 53. The stomata are arranged quite regularly and look like round openings with double sides. These sides close up partly across the opening in dry weather and so prevent the water vapour from passing out too quickly. In spite of them there are days in the summer when the leaves get rid of the water more quickly than the root can suck it from the parched ground. Then the leaves begin to flag and hang limply down until the falling dew freshens them up again. In damp weather the stomata open, as when the air outside is moist evaporation will in any case be slow.

Fig. 54. Leaf of Guelder Rose, showing blade, stalk, and stipules.

A typical leaf such as that of the plum or guelder rose has two parts besides the blade, namely a thin flexible stalk, and at the base of the stalk two small green leaves known as **stipules** (Fig. 54). These stipules vary greatly in size in different kinds of leaves. In some they are larger than the blade itself, in others they are either altogether absent or drop off almost as soon as the leaf comes out of the bud. In this last case they often take the form of scales which shelter the young leaves before they are fully grown. If you look at the leaf bud of a cherry or horse-chestnut you will there see the stipules going through every stage from scales to small green leaflets.

I said that the stalk is flexible. What do you think would happen if it were not? The first storm of rain or wind would strike against the flat surface of the leaf and either tear it to pieces or snap the stalk in two. It is usually long as well, so as to bring the leaf out from the shade of the other leaves and expose it more fully to the light.

PRACTICAL WORK.

1. You are given two laurel or ivy leaves and two softer leaves (*e.g.* lilac) all plucked straight from the twigs. Peel a piece of outer skin off one of each pair of leaves. Place all four leaves in the sun. Later in the morning go and look at them and write down the answers to these two questions :—In what order do the leaves begin to fade ? What is the reason of this ?

2. (**Demonstration by Teacher.**) Two similar twigs covered with leaves (a kind with good sized leaves is best) are placed in specimen glasses full of water and the opening is plugged with cotton wool. The leaves are then all stripped from one twig. The glasses, with the twigs in, are placed on rough balances (the small ones used for weighing letters will do), and are counterpoised with weights and shot or pebbles. Both balances are now placed in the sun, still bearing the glasses and shot. They are examined at the end of $\frac{1}{2}$ hr. and again at the end of 1 hr. Illustrate this experiment by diagrams. Underneath your pictures answer these questions : (*a*) Did you see any change at the end of $\frac{1}{2}$ hr. or 1 hr. ? (*b*) How do you account for this ? (*c*) What does the experiment show you ?

3. Fold your paper in four columns, and head them with the words Name, Blade, Stem, Stipules. Examine the leaves of ivy,

hawthorn, elm, honeysuckle, rose or blackberry, apple or pear, and according as you find each part present or not write "yes" or "no" in the column belonging to it.

4. With a dissecting needle (or a long pin) take all the scales off your cherry or horse-chestnut bud. Arrange them in order on a piece of paper and leave for inspection.

HOME WORK.

Collect at least six different kinds of leaves with well-marked outlines and veins. Lay each leaf as nearly flat as you can on paper. Draw its outline in pencil and put its name beneath. Go over the outline of each leaf very carefully in ink correcting all mistakes and drawing a clear outline. Place your leaves between sheets of botanical paper or blotting paper, and leave them under some heavy books to press until the lesson after next.

Lesson 2. THE ARRANGEMENT OF LEAVES.

Season. Second week in May.

Materials required for each pupil.

Small spray of eucalyptus showing three or four leaves growing. (See Appendix III.)

One leafy twig of each of the following:

Aspen (or poplar), ivy, yew, horse-chestnut, lime, birch, ash, oak, walnut, elder, osier willow. Pressed ivy leaves of various sizes.

Quarter plate printing frame and glass to fit it. Two or three sheets of ferro-prussiate printing paper (to be obtained to order at any chemist's, but should be used fresh).

Pair of compasses. Ruler measuring tenths of an inch.

The leaves on our native trees always grow from the twig with their flat side facing the sky. The reason for this is that in a country as far north of the equator as ours is they want as much light as they can get, and they get more in that position than in any other. In countries, however, like Australia, where the sky is without a cloud for months together, the leaf blades grow twisted on their stalks, so that only the thin edge is towards the sky and the flat surface does not get the sun's rays beating directly on it.

You have probably noticed the perpetual trembling of the leaves of the aspen tree and have been told several pretty legends to account for this. The real reason is only that the stalk is rather pinched from side to side where it joins the leaf blade, with the result that the blade is fastened on less stiffly than in other trees and sways to and fro, even when there is no perceptible wind to move it.

In order to appreciate the part played by leaf stalks in spreading out each blade to the light, you should look at an ivy spray that has grown against a wall. Here the leaves could not grow in their usual position, with their flat sides facing the sky, because the wall got in the way. To make up for this in some cases a stalk has grown longer than the others, in others it has twisted round so as to place the blade in a better position. The result is that the lobes of each leaf fit in between those of the ones next to it in a wonderful way and form what is known as a **leaf mosaic**. This name is given to the pattern because the effect is like the mosaic pictures Italians make by fitting together small pieces of stone. This same thing is seen in many elm twigs, and some botanists think that the object of the large and small side of this kind of leaf is to allow the separate leaves to fit closely together (Fig. 55). A twig of maple looked at from above shows the same thing. The lower leaves, instead of being completely overshadowed by the upper ones, are furnished with longer stalks, which enable them also to catch the sunlight from above.

In order to grow well some kinds of trees want more light than others do, and if other trees grow too closely round the light-loving ones those branches that are

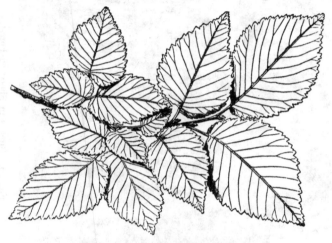

Fig. 55.   Elm twig, showing leaf mosaic.

overshadowed will not grow as freely as those that are exposed to the sun and air.   It has been found that the shade-bearing trees are those which, like the beech, themselves cast the deepest shade, while the trees with light crowns will not grow under more thickly leaved ones.   This is probably because in a perfectly natural state trees grow in a clump with others of their

own kind.   Thus the feathery birch would not mind the
sun-flecked shadow of other birches.

Fig. 56.   Birches in Sherwood Forest.

When you looked at leaf buds you saw that leaves
are not scattered anyhow on the twigs from which they
grow, but are arranged on it at definite places called

nodes the distance between which varies with the kind
of tree.   Leaves, like branches and flowers, grow either

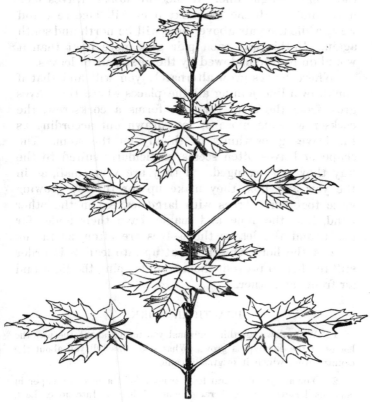

Fig. 57.   Shoot of Norway Maple, showing how the leaves grow so as to
get as much light as possible.

opposite each other or alternately.   When the arrangement is opposite it will usually be found that each pair
of leaves grows not directly above the pair below it but

G.   8 at right angles to it (Fig. 57). Thus the leaves grow up the stem in four vertical lines with every other leaf in the same straight line. If the two lowest leaves face north and south the two next ones will face east and west, while the pair above them will be north and south again. In this way each pair gets more light than it would do if overshadowed by the next pair of leaves.

Where leaves grow alternately you will find that if you draw a line joining all the places where the leaves grow from the stem this line forms a corkscrew, the corkscrew being more or less drawn out according as the leaves grow thinly or thickly on the stem. The shape of leaves often seems particularly suited to the way they are arranged. If they are very small, as in the yew and box, they make up for this by growing close together. Trees with large leaves, on the other hand, like the lime and maple, have their nodes far apart, and the leaves themselves are often, as in the case of the horse-chestnut, cut up into leaflets in order still further to prevent them from keeping the light and air from each other.

PRACTICAL WORK.

1. Write down anything unusual you notice about the way the leaves of the eucalyptus grow. What does this tell you about the country from which it originally came ?

2. You are given several leafy twigs. Fold a piece of paper in columns headed Name, Arrangement of leaves, Internode, Leaf. Now look at your twigs and in the case of each write down the name of the twig, whether the leaves are opposite or alternate, the length of each internode (in inches and tenths) and the length of a leaf from one end to another.

3. Taking each of the twigs with alternate leaves, draw on it with a pen or compass point a line connecting the points of insertion

of the leaves. Write down the number of leaves (counting inclusively) that lie on this line in passing from one leaf to the next exactly above it.

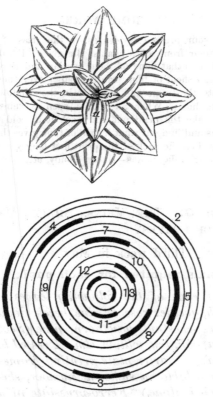

Fig. 58. The arrangement of the leaves of a Plantain (showing how question **4** in the practical work is to be done).

4. Draw with your compasses a number of circles one inside the other (make the largest about 2 in. in diameter, the next 1¾ and so on). Now take a piece of one of your twigs that has as many leaves as you have drawn circles. Stand the twig up on end in the middle of your circles and mark on the largest circle the position of the lowest leaf, on the next circle the next leaf, and so on. Leave paper and twig for inspection. (Fig. 58 gives an example.)

HOME WORK.

Arrange your pressed ivy leaves on the glass of the printing frame with their front sides against the glass. Fit different sized leaves together so that none of them overlap and they make a nice pattern. Put a sheet of blue printing paper over them. Then put the back into the frame and stand it in the window. Look at it from hour to hour. When all details, such as veins, have nearly disappeared take the print out and wash it in cold water. Make several prints and find out what length of time gives the best results. Write in pencil on the back of the print how long you were printing it and what sort of day it was—sunny, cloudy, etc.

Lesson 3. THE VEINS, SHAPE, MARGIN AND SURFACE OF LEAVES.

Season. Third week in May.

Materials required for each pupil.

One of each of the following leaves:—Iris, lime, maple, ash, guelder rose or vine, beech, dogwood, black-berry, horse-chestnut, hawthorn, pine (Scotch), rowan or robinia, oak, ivy, willow, elm. Pressed leaves (from first lesson). Quarter plate printing frames. (If these are not available in sufficient number, sheets of glass may be tried alone.) Ferro-prussiate printing paper to fit the frame. Basins of water for washing the prints.

The veins of a leaf, besides strengthening the thin blade, act very much as our veins do, that is they supply every part of the leaf with sap. Every vein, even the smallest one, consists of two pipes running side by side, one of which brings up the sap while the other carries away the starch as soon as it is made. Hold up any ordinary leaf such as a pear leaf to the light and notice how the veins branch and rebranch until you can hardly see them. In all our British trees the veins form a network very similar to this one,

Fig. 59. Leaf of Birch, showing **pinnate** venation.

although, as you will see directly, the arrangement of the larger veins may vary. Now compare this leaf with an iris or grass leaf. In these last the veins are all arranged side by side and do not branch at all. This kind of veining is called **parallel** and is found in many tropical leaves. If you look at the veins of birch or lime and currant or maple leaves you will find that

although they are both netted there is one great difference between them. In the first case there is one
obvious principal vein or mid-rib from which smaller
veins branch (Fig. 59). In the second there are several
veins of equal size all springing from the place where

Fig. 60. Leaf of Red Currant, showing **palmate** venation.

the leaf blade joins the stalk (Fig. 60). These veins
have been compared to the fingers on a hand and have
caused the name of **palmate** venation to be given to
this type. The lime shows an example of the more
common **pinnate** venation (so called from a Latin word
pinna, meaning a feather). All our native trees show

one of these two venations, although there may be differences in the courses the veins run and the amount of branching they show.

The leaves themselves vary so much in shape on different trees that you can make a very interesting collection by seeing how many different shapes of leaves you can find. In the first place some leaves are all in one piece or **simple**, while others are made up of a number of leaflets, that is they are **compound**. Then they show every gradation in shape from the needle-shaped leaves of the pine to nearly round hazel leaves. Leaves in between these two extremes may be lance-shaped like the privet, oblong, or oval. Other leaves again may be heart- or arrow-shaped. These names will, however, mean nothing to you until you know the leaves themselves and try to describe their shapes. In order to give a clear word picture when a leaf is being described the shape, venation, margin, and surface should each be taken in turn. We have spoken of the first two and will now pass on to the margin and surface.

A leaf's margin is often quite smooth as in the beech, when it is said to be **entire**. Other margins are made up of tiny teeth like a saw, or they may be **wavy** or **lobed**, or may cut the leaf blade up until the leaf is nearly a compound one. The shape of some leaves, which like those of the ash are drawn out into long points called "drip tips," makes the rain drops run quickly off the leaves, and some botanists think that the trembling of an aspen leaf serves the purpose of shaking off the rain. Be this as it may there is no doubt that the surface of the blade usually shows some device to ensure that water shall not remain there long, for although most of the stomata are on the under side of the blade there are generally enough on the top to make it important that they should not be choked with rain.

Fig. 61.   Leaf of Ash, a compound leaf with the ends of the leaflets forming " drip tips."

Sometimes the leaf is smooth and shiny, like a laurel leaf, or again it may be covered with a waxy coat or with hairs.

These differences are only to be found on the upper surface, as the under one does not run the same risk.

An advantage of learning some of the words like palmate, pinnate, simple, compound, entire, wavy, lobed, which are printed in black type in this lesson, is that you can then make one word do instead of a dozen when you try to describe some new leaf.

PRACTICAL WORK.

1. Place one of your pressed leaves (if unpressed leaves are used their dampness may spoil the printing paper) in the printing frame with the sensitive paper. Make the print just as you did in your last home work, but leave it just long enough to get a good impression of the veins. Wash it and put it on one side to dry. While it is printing go on with the other questions.

2. Trace the outlines of the leaves supplied, *e.g.* lime, guelder rose or vine, iris, birch, dogwood, blackberry, and finish in ink.

3. Write down which of the following leaves are simple and which compound, and if compound the number of leaflets :—horse-chestnut, hawthorn, maple, pine, blackberry, rowan or robinia.

4. Draw each of the following leaves and describe exactly, but in as few words as possible, their shape, venation, margin, and surface :—ash, oak, maple, ivy, willow, elm.

HOME WORK.

Draw a twig with leaves in natural position, attending to all you have learnt about the shape of the leaves, their arrangement on the stem, etc. Finish in ink.

Lesson 4. THE MODIFICATION OF LEAVES.

Season. Fourth week in May.

Materials required. (*a*) **For demonstration.**

A cactus plant in pot (see Appendix III).

(*b*) **For each pupil.**

Twigs of broom, cypress, gorse, and butcher's broom. Pansy leaf. Part of a house leek showing leaves (one of the thick-leaved sedums may be substituted).

One spray of each of the following: common vetch (Vicia sativa), yellow vetchling (Lathyrus Aphaca), Smilax (from any nurseryman).

One of Nature's most universal laws, which as you grow older you will find at work all round you, decrees that every living thing shall learn to suit itself to its surroundings. If it cannot do so it must make way for those that can. Green plants are found growing in such very different surroundings, from the wind-swept plains of Siberia to tropical swamps, that they naturally take many forms ; and of all the parts of which they are made up no part changes as much with its surroundings as does the leaf.

The large flat surface of the leaf blade is very useful to a plant as long as its roots can absorb as much water as it needs to make up for the quantities given off by the leaves. Some plants, however, grow under such circumstances that, either because there is not much water to be had or for some other reason, **transpiration**, *i.e.* the giving off of water, must be lessened. This can be done in several ways. Sometimes the leaves are crowded closely together or have a waxy or hairy coat so that the air cannot reach them freely, but more frequently the surface of the leaf is diminished. Besides the pines and firs of which mention was made in an earlier chapter the heaths show a good example of this reduction of surface. They live on moors and exposed places where if they had broad leaves the cold dry winds which often sweep over them would make

Fig. 62. Black Wood of Rannoch, Perthshire, showing Scotch Pine and Ling.

evaporation go on too quickly and would chill the plants just as you chill your hand if you blow on it when it is wet. This is prevented by the leaf surface being curled over towards the mid-rib, so that the stomata which are on the under side are protected from the wind and can give off moisture as gradually as if the plant grew in a more sheltered position.

When plants grow on the sea shore, or in marshes

where there is a good deal of salt, all the water they suck up contains salt dissolved in it. When the water vapour passes out through the leaves the salt remains behind, hence such plants must be able to get on with as little water as possible. That is why most grasses and reeds found by the sea have their long leaves closely rolled up. When the air is moist so that there is no danger of evaporation going on too quickly, some of them unroll themselves, and curl up again in dry weather.

Many desert plants live in a climate where, not only is the *air* dry, but during a portion of the year there is no moisture at all to be got from the ground. Such a plant must do one of two things : it can die down to the root during the drought, doing all its rapid growing in the rainy season ; or it can live through the dry weather on a store of water which it put on one side when there was plenty to be had. The leaves or sometimes the stems of such plants become thick and fleshy and act as storehouses. This is seen in members of the cactus family, and to a smaller extent in some of our stone-crops which are accustomed to live on rocky ground where the water supply is not to be depended upon.

In our country there is plenty of water to be had as a rule, but the length and coldness of our winters makes it impossible for plants to go on actively growing through it, hence most of our British trees cast their leaves before the winter comes, and do not begin to sprout again until the warmer weather. Those leaves that are evergreen are furnished with tough leathery coats which prevent evaporation from going on quickly and protect the inside of the leaf from cold winds.

Although most ordinary leaves have a large blade and a thin stalk with two small stipules at the end, each of these parts may in some plants and trees look so different that you would never recognise them. In the pansy, for instance, the stipules have grown as large as the blade. In some climbing plants the stipules have turned into long curling tendrils, while in others the blade has become a tendril and the stipules are large enough to take its place. The position in which these parts grow help to tell us what they really are, and botanists find out more about them by comparing plants belonging to the same family, or sometimes by noticing the stages they go through in growing from little seedlings to full grown plants. In the robinia or false acacia two strong spines grow from the base of each leaf. The place where they grow shows that they are really stipules. They have turned into spines to keep off any animals that want to feed on the tender young shoots. One of the robinia's foreign cousins is still better protected, for its stipule spines are large and hollow, and are inhabited by a fierce kind of ant which attacks any enemy that ventures too near.

In the case of the gorse not only the stipules but the very leaves themselves and the branches have turned to spines. How do we know that all the spines with which a gorse bush is covered are really leaves and branches? Well, in two ways. Firstly by noticing the relative position of the spines and the buds or flowers. The buds are borne *in the axils* of some spines and *upon* others. Therefore some of the spines are leaves and others branches. Secondly by looking at a very young gorse seedling (Fig. 63). We see that at first it is provided with soft leaves not unlike those

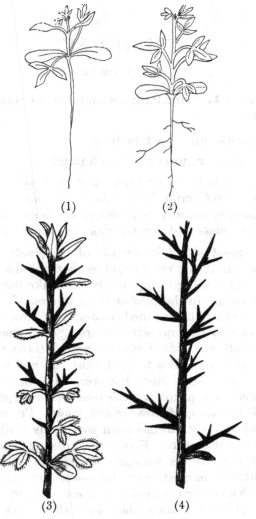

Fig. 63. Stages in the development of Gorse thorns.
(1) and (2) Gorse seedlings. (3) and (4) Twigs at later stages.

CHAPTER XV

THE FLOWER

Lesson 1. THE PARTS WHICH GO TO MAKE UP A FLOWER.

Season. First week in June.

Materials required for each pupil.

One whole bunch of horse-chestnut flowers. One of each of the following flowers: hawthorn, wild hyacinth, laburnum, sycamore (one male and one hermaphrodite), beech (one male and one female).

The great object in the life of all plants, if their kind is not to die out altogether, is to produce young plants which will grow up and in time take the place of the old ones. In the case of trees this may be done in several ways. Shoots called **suckers** may grow out from the root and become separate trees, or if small twigs are cut off and placed in the ground these **cuttings** will send out rootlets from the cut end and will grow into young trees. Just, however, as with nearly all other classes of plants, a tree's usual way of reproducing itself is by means of **flowers** and **seeds.** Perhaps until you looked at catkins last term you hardly realised that trees do bear flowers. Every tree, growing in a climate that suits it, flowers sometime during the course of its life, although in most cases the flowers are so small and inconspicuous that, unless you knew what to look for, you might never know they were there. As the life of a tree is so long, there is no need for it to flower

every year like a small plant that has only a few years
to live ; hence most trees do not bear flowers until
they are twenty or thirty years old, and some slow-
growing trees like the oak wait longer still.

In order to see the different parts of which most
flowers are made up we will look at a single horse-
chestnut flower, taking it from the lower part of the
bunch (Fig. 64). You see that the flower consists of
four rings of different kinds. First a cup called the
calyx which covered up the flower completely when it

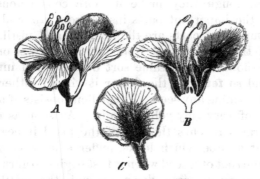

Fig. 64. *A*. Horse-chestnut flower. *B*. Same flower cut in two, showing
pistil in centre. *C*. Single petal.

was a bud and is composed of green **sepals**, not unlike
little leaves. Then come the coloured **petals** which
together form the **corolla** and are the showy part of
the flower. If you hold one of these up to the light
you will see how delicately veined it is. When the
petals are pulled off, the next layer is seen to be a
number of long **stamens**, with yellow pollen covering
their heads. These heads are really little boxes made
to hold the pollen. Right in the middle of the flower

is the **pistil.** Its lower end is swollen to form the **ovary** and contains the **ovule**, a little round green body, which will later on ripen into the seed, the horse-chestnut you know so well.

Most ordinary flowers show these four parts, although the numbers in each ring vary a great deal according to the kind of flower. In some flowers, however, you will find that the calyx or the corolla is missing, and even sometimes, as in the catkins, that neither is there. Neither calyx nor corolla is really necessary to the flower although they make it more conspicuous and protect the important parts inside from rain and wind. These important parts are the stamens and pistil, for if some of the pollen either from the same flower or from another flower of the same sort does not fall into the pistil and so **fertilise** the flower it will all wither away without forming any seed. The pollen consists of a large number of very tiny yellow grains. As soon as one of these grains reaches the top of the pistil it begins to send out a tube which looks, under a microscope, not unlike the root of a seed when it first begins to germinate. This tube grows right down through the pistil until it touches the ovule. This, as it were, wakes the ovule up, and it begins to swell together with the ovary surrounding it, until in time the ovule becomes the seed and the ovary the fruit. Sometimes if we have very cold weather when the fruit trees are in flower the fruit instead of "setting" falls off as soon as the flowers are faded. What has happened is this: the frosty weather has nipped the delicate pollen tubes just as they were growing and has killed them. Hence the flowers are not fertilised and the fruit will not set.

Since pollen from another flower of the same kind

does quite as well as that from the same flower, stamens and pistil sometimes grow separately as in the case of the catkins. When this is the case some arrangement is necessary to ensure that the pollen shall be carried to the pistil.

PRACTICAL WORK.

1. Draw exactly 10 times their natural size :

   (*a*) A flower from the top of your horse-chestnut bunch.

   (*b*) A flower from the bottom of it.

2. Describe as shortly as you can any difference you see between the two flowers.

3. Take a flower carefully to pieces with a dissecting needle. Are the petals joined together at their base or do they come off separately ? Are the sepals joined at their base ?

4. Draw exactly five times their natural size : (*a*) a stamen from a newly opened flower, (*b*) a stamen from a flower that is beginning to fade.

5. Are all the stamen heads alike ? What differences do you find ? Are any of them dusty with pollen ?

6. Draw a pistil exactly five times its natural size.

7. Have all the flowers pistils ? How do you distinguish the pistil from the stamens ? What do you see at the base of the pistil ?

8. Strip the bunch of its flowers and draw (natural size) the whole stalk with the side stalks growing from it.

9. Dissect one of each kind of flower given to you, both of the beech and sycamore flowers. Arrange each flower on a separate piece of paper, putting all the sepals in one corner, the petals in another, and so on. Leave them for inspection.

HOME WORK.

Pick twelve flowers, all of different kinds. Take a piece of paper and fold it in five columns. Head them as follows : Name of flower,

Sepals, Petals, Stamens, Pistil. Now write the name of each flower in the first column, and after carefully dissecting the flower, write the number of sepals, etc. in the columns to which they belong. In the case of trees with two kinds of flowers do one of each kind.

Lesson 2. FOLIAGE LEAVES AND FLOWER LEAVES.

Season. Second week in June.

Materials required for each pupil.

One wild rose. One garden rose. One fern leaf with spores. Pine twig with male flowers and three stages of cones. Pea pod and leaf.

When we looked at the horse-chestnut flowers we noticed that the sepals and petals were in some ways not unlike leaves. The wild rose shows this likeness better still, as the green sepals grow very long and branched (Fig. 65). In the case of some flowers the stamens or petals look exactly like leaves. This is seen in the case of badly formed garden roses which, instead of being made up of row within row of petals, have two or three rows only of ordinary petals and a little bunch of green leaves in the middle. These examples show that flowers and leaves are not really very distantly related. It is probable that a very long time ago in the history of plant life there was much less distinction between them, the only difference at first being that on some leaves grew pollen and seeds. (This is still seen in ferns, which bear no flowers at all, and, as I will show you directly, in fir cones.) Later on some of the leaves began to be folded up so as to form little boxes, some to hold the pollen and others to hold the seed. In each generation that followed these special pollen and seed bearing leaves became more and more like the stamens and pistils we see now. At the same time other leaves began to grow round these and grew more brightly

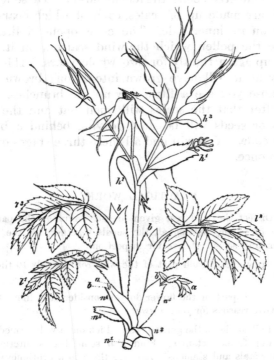

Fig. 65. Inflorescence of a Rose, showing bud scales *n* below the ordinary foliage leaves *l* ; above these are **bracts**, *h*, from the axils of which the flowers grow.

coloured so that at last flowers as we now know them were developed.

A case in which the pollen and seed is still borne uncovered on flat scales is seen in all trees of the pine family. If you examine a Scotch pine in late May or June you will see that some of the twigs end with bunches of male flowers covered with pollen, and others with a little green cone, the female flower. These female flowers are made up of scales, each of which bears two ovules on its inner side. The cone opens a little to receive the pollen which the wind scatters on it, then closes up again tightly for two whole years. This time next year it will have grown into a long brown cone like those you see a little way up the branches. The year after that the scales will open out and the ripe ovules or seeds will be shed, leaving behind a brown empty cone. You will find all these three stages on the tree at once.

PRACTICAL WORK.

1. Dissect the wild rose given to you. Arrange the parts on a piece of paper, putting all the sepals together, all the petals together, and so on. Leave for inspection.

2. Draw one of each part of the wild rose exactly 10 times its natural size.

3. Which part of the flower do you consider most like a foliage leaf? Give reasons for your answer.

4. Pull to pieces the garden rose. Pick out a well formed petal and a well formed stamen. Find also some leaves intermediate between petals and stamens. Arrange them in a continuous series with the petal at one end and the stamen at the other end. Leave for inspection.

5. Draw the back and the front of your fern leaf side by side exactly its natural size.

6. Describe what you see on the back of the fern but do not see on the front. What happens if you scrape them with the point of your compasses? What do you think they are?

7. How many kinds of cones do you see on your pine twig?

Draw the smallest cone four times its natural size and the others exactly their natural size and write by the side of each cone the date of the summer when it began to grow.

8. What do you see on the twig besides leaves and cones? Whereabouts on the twig do they grow? What happens if you shake them?

9. Take several scales from each kind of cone. What differences do you see in them? What has happened in the case of the scales from the oldest cone?

HOME WORK.

Fold a pea leaflet in two and place it by your pea pod in such a position that the two look as much alike as possible.

Draw the leaflet and the pod the same size and mark in your drawing the side of the pod which you think corresponds to the mid-rib of the leaf. Open the leaflet again and split your pod along its shortest side. Now draw them both open and mark in the drawing the part of the leaflet which corresponds to the line along which the peas grow in the pod.

Lesson 3. FLOWER PATTERNS.

Season. Third week in June.

Materials required for each pupil.

8—10 *small paper labels with stout cotton for tying. One head of each of the flowers: speedwell, wheat, arabis or wallflower, red campion (branching at least twice), privet, cow parsnip, elder, plantain.*

Flower branches follow definite rules just as leaf branches do, and as you look at them you will notice many likenesses between them. In the first place a flower branch always grows out of the axil of a leaf.

This leaf often differs in shape from the ordinary leaves, and is called a **bract**. Sometimes it falls off early so that you will not find it, and sometimes, as you will see later on, it does not grow where you expect it to, but its natural position is always the same as in the case of

Fig. 66. Barberry flowers.

a leaf branch. The flowering part of a plant is known as the **inflorescence** and takes many forms, varying from cases as simple as the tulip in which each stalk sent up from the bulb ends in a single flower, to those like the horse-chestnut or elder which branch several times.

The first large class of inflorescences contains all those in which the main stem, giving off flowers from its sides, goes on growing all the time so that its growth is **indefinite**, *i.e.* unlimited in length. The oldest flowers

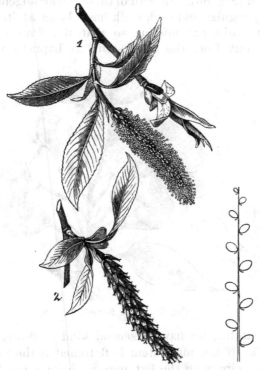

Fig. 67. (1) Male and (2) female Willow catkins.

will thus be those at the base of the inflorescence, and the young ones further up, with the smallest buds at the very top. These indefinite inflorescences depend for their shape on the length and arrangement of the small flower stalks on the main stem. The simplest form is seen when these stalks are all about the same length and are arranged alternately all the way up the main stem. This form is seen in the barberry and the laburnum (Fig. 66). In both of these the arrangement is perfectly regular and each stalk has a bract at its base. If these stalks are missing so that the flowers grow straight out from the main stem, as happens in the

Fig. 68. Pear flowers.

willow catkins, we have a second kind of flower head (Fig. 67). If the main stem is flattened at the top and the flowers grow on the flat part we have a result like the cornflower, in which a single head is made up of a number of small flowers. Sometimes, again, we see a form in which the stalks are there, and are arranged alternately as in the barberry, but, instead of all being the same length, the lower stalks grow a good deal longer than the upper ones so that the flowers are all at about the same level. This case is seen in the pear (Fig. 68). Now suppose that the spaces between the flower stalks disappear altogether so that the stalks all spring from the same level like the ribs of an umbrella. You have now the form of flower head that is seen in

Fig. 69. Ivy flowers.

ivy and hedge parsley, and many other wild flowers (Fig. 69). If you imagine that this form is really a long spray like the laburnum which has been pressed together like a telescope you see that the young flowers that were then at the top will now be down in the middle. We have now spoken of five forms of flower head. One with alternate stalks, two with no stalks at all, one with a partly flattened head and one with a quite flattened head. Each of these inflorescences may be

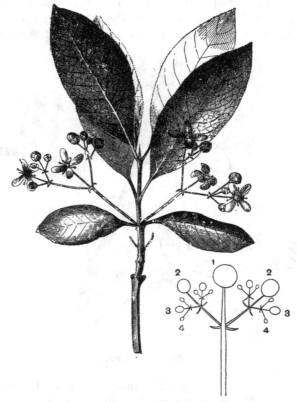

Fig. 70. Flowers of Spindle Tree.

complicated by the flower stalks branching again and thus themselves belonging to any one of these groups.

The second main class to which inflorescences may

belong is that in which the growth of the main stem is
definite, that is to say it stops growing after a time and
growth is carried on by the side branches. If two stalks
grow out opposite each other at each node this results
in a very regular type such as you see in the buttercup
or in the spindle tree (Fig. 70). If you imagine this
squeezed flat you will get something that looks at first
like the same shaped flower head as the hedge parsley.

Fig. 71. Flowers of Laurustinus.

It cannot, however, really be the same because the
young flowers instead of being in the middle are found
dotted all about (Fig. 71).

Suppose now that only one branch grows from each
node the result is a zigzag. When this is straightened
out it all looks like one stem. What is the one difference
which shows that this cannot be an inflorescence of
indefinite growth like the laburnum? The bracts, instead

of growing at the base of each leaf stalk, always spring from the opposite side of the stem. If therefore you see an inflorescence that looks as if it grew exactly like the laburnum except that the bracts have apparently got shifted, you will know that it must really belong to this last class and that all the separate parts of which it is made up have been straightened out to look like one stem. Without these bracts you would find it impossible to tell to which of these classes a flower belonged.

PRACTICAL WORK.

1. Draw diagrams (life size) of the following inflorescences : speedwell, plantain, wheat, arabis or wall-flower, cow parsnip.

2. In the case of each of the above flowers do the stalklets grow opposite each other or alternately ? are there any bracts present ? whereabouts are the youngest flowers ?

(Write the name of each flower on a new line with the answers to the questions by its side.)

3. Draw your red campion plant life size and make a diagram of its branching at the side.

4. Do the same number of branches grow out from each node ? if so, how many ? Does each branch go on growing indefinitely or does it stop at a special point ?

5. Tie one of your labels on to each campion flower marking the oldest " 1," the next "2," and so on.

6. Draw a diagram (twice life size) of the arrangement of flowers in a head of privet. Draw small circles to represent the flowers themselves, making the largest circles for the oldest flowers.

7. Do the bracts grow from the base of the side stalks or on the opposite side of the main stem ?

8. Whereabouts are the youngest flowers in the cow parsnip ? Whereabouts are they in the elder ? How do you account for this ? Illustrate your answer with diagrams of both inflorescences.

HOME WORK.

Design some patterns suitable for embroidery, for wall papers, or for book covers, based on flowers. Let the general style and pattern of branching be true to nature. Show some flowers in bud, some full open and others in seed ; and in doing so show that you know which parts of the inflorescence will flower first and which last.

Lesson 4. THE FERTILISATION OF FLOWERS.

Season. Third week in June.

Materials required for each pupil.

One head of each of the following flowers : white dead nettle, stinging nettle, honeysuckle, lime, flowering grass, elder, rosebay, sage.

A single flower of each of the following : pansy, foxglove, nasturtium, sweet pea.

When a flower contains both stamens and pistil it is obvious that the simplest way in which it can be fertilised is for pollen from the stamens to fall straight on to the pistil. In many cases this self-fertilisation, as it is called, is what occurs. If this, however, goes on for many generations, and the flowers never get any pollen from another plant, they tend to produce fewer seeds, and the young plants which grow from these seeds become weakly. Nature has, therefore, several ways of carrying pollen about from one plant to another and so cross-fertilising them. Her two chief agents are insects and the wind, and flowers are adapted to whichever of these is going to be used.

All the garden flowers you know best are fertilised by **insects** such as bees, butterflies, moths, or flies, which are attracted to the flowers by their bright colour and in many cases by their scent. If this was all the flowers had to offer insects would soon learn that no food was to be had there and would not come again. Food is, however, supplied in the shape of pollen which bees use to make bread for their young, and, besides this, many of these flowers bear a store of honey right down inside the corolla. Insects come for the honey, and, while they are sucking it, brush against the stamens, and carry away some of the pollen dust on their heads when they go. When they visit the next flower some of this pollen will probably be knocked off on to the top of the pistil which is made sticky on purpose to hold it. In this way insects of all sizes, or even, in the case of tropical flowers, tiny humming birds, fly from flower to flower unconsciously fertilising them in return for the honey they give.

The **wind** is, of course, a far less certain fertilising agent than are these insects. It will blow the pollen off flowers and will carry it away in the air and scatter it in all directions. Therefore besides being very light, the pollen of such flowers must be produced in enormous quantities, for, although some of it will fall on other flowers, a great deal will be wasted. Pine trees show this very well, and in some places where forests of them grow such a cloud of pollen falls in spring as quite to colour the ground and make the country people think there has been a sulphur shower. In flowers fertilised by the wind the pistil is often branched or hairy so that it will hold the pollen that is blown on to it, and the calyx and brightly coloured corolla which in our first class of flowers served to protect the pollen and to attract insect visitors, are found to be absent. They would be not only useless but in the way, since they would keep the wind off. That is why many tree flowers are reduced to mere bunches of stamens and pistils dangling, in the shape of long catkins, in the wind.

Flowers show many different devices, the object of which is to prevent them from self-fertilising themselves. The simplest and most common one is seen when, as in the wallflower, the pistil stands up above the stamens. It is thus the first thing to touch any bee that visits the flower, and if some pollen is already on the bee this will reach the pistil before its own pollen does.

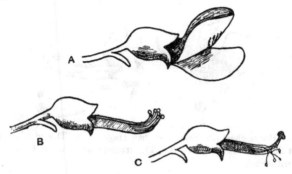

Fig. 72. *A*. Laburnum flower. *B*. The stamens of a young Laburnum flower before the pistil is ripe. *C*. The pistil of an older flower with the stamens withered.

Another arrangement very often found is for the stamens and pistil to ripen at different times, that is to say, when the pollen boxes open and the pollen is ready for scattering, the pistil is still too young to receive it; later on, when the pistil is full grown, the stamens are all withered, and any pollen to be of use must come from some other flowers in which the stamens are at an earlier stage. You will understand this process better by looking at a laburnum flower (Fig. 72): a bee alights on the flower in search of honey and the weight of his body on the point of the lower petal presses it down and releases the stamens and pistil which were shut up inside it. The stamens spring out and dust all the under side of the bee's body with pollen. The pistil is still quite young, and must go on growing before it is ready to receive pollen grains. The bee now flies on to an older flower. Its stamens are old and have lost all their pollen, but the pistil has grown long, and, as it brushes against the under side of the bee where the pollen from

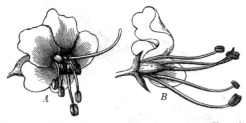

Fig. 73. Horse-chestnut flowers in *two* stages. *A*. The pistil is ripe but the stamens are not. *B*. The stamens are ripe and the pistil is withering.

the first flower is, some of the pollen grains stick to the pistil and thus cross-fertilise the flower.

A case that is also found, although not as often as this last one, is when the pistil is ready before the pollen boxes are open, so that to fertilise the flower a bee must have first been to an older flower in which the pollen is ready for distribution. If you examine horse-chestnut flowers in different stages you will see that in the younger ones the pistil sweeps upwards so that it rubs against any bee that visits the flower, while the stamens hang down out of the way (Fig. 73). In older flowers the stamens have grown up till they are in very

much the same position as the pistil was before, while
the pistil has now withered away. You see that in
either the laburnum or the horse-chestnut it is impos-

Fig. 74.   Honeysuckle flowers, showing the long tubes adapted to
fertilisation by moths.

sible for self-fertilisation to take place.   A large number
of flowers belong to one of these two classes.

Other flowers show either hinged stamens which
when touched at the bottom move round and brush

against the visiting bee, or some explosive mechanism which makes them when touched jump up and scatter the pollen. If you notice flowers carefully and watch the various insects gathering honey, you will see what an interesting subject this is, and how each flower is suited to the size and shape of the particular insect that is found there, as the foxglove to humble bees, honeysuckle to long-tongued hawk-moths, and night flowers like the evening primrose to night-flying moths.

PRACTICAL WORK.

1. Pull the petals off your dead nettle flowers. What do you notice near the base of the pistil? Has it any taste? What is it? Why is it there?

2. Draw (natural size) the flowers of pansy, nasturtium, fox-glove.

3. Do you see any spots or stripes on the petals of any of the flowers? If so which way do they lead? Why are they there?

4. Some trees with wind-fertilised flowers, as the ash and hazel, flower in early spring. Can you suggest a reason for this?

5. You are given flowers of stinging nettle, honeysuckle, lime, flowering grass, elder. Head five columns as follows: Name, Colour, Scent, Honey, Fertilised by insects. Now look carefully at each flower and write down its name and colour in the first two columns and yes or no in the last three.

6. Draw a sweet pea flower (four times natural size) front and side view. Mark with a cross in both drawings the exact place where you think a bee would alight on the flower.

7. Taking the flower itself press with your pencil point where the bee would settle. What happens? What happens when you take the pencil away?

8. Draw (four times natural size) a flower from near the top of your rosebay spike and one from near the bottom.

9. What are the chief differences in the two rosebay flowers? After examining the various flowers on your spike write a short sketch of the life of a single flower showing the order in which the different parts open. Is it possible for such a flower to fertilise itself? Give a reason for your answer.

10. Take one flower from near the top of your head of sage and one from near the bottom. What differences do you see between them? Touch with your pencil point the lower end of the stamens in the younger flower (the one from the top). What happens? What will happen to a bee looking for honey in this flower? Suppose it flies next to an older flower where will it settle? Will the pistil or stamens touch it first? What will happen to any pollen sticking to its back?

11. Remove the stamens from the younger flower of sage. Draw them (five times natural size) in both positions.

Remove the pistils from both flowers and draw them (five times natural size).

12. Out-of-doors. Watch the flowers of dead nettle, pansy, nasturtium, foxglove, honeysuckle, lime, elder, sweet pea, rosebay and sage. Write a list of the insects you see visiting each.

Provide yourself with a test tube and plug of cotton wool. Catch a bee whilst it is busy on the flowers of the lime. Is there any pollen on the bee, if so, on what part of it?

HOME WORK.

1. Draw any five kinds of flowers each showing one of the devices mentioned in this lesson and say what the device is.

2. Write a list of any insects you see visiting the flowers of horse-chestnut, lilac, syringa, ivy, holly, pear, lime, plum, or other trees.

APPENDIX I

BOOKS

The teacher will find the following books useful :

AVEBURY, LORD, *Notes on the Life History of British Flowering Plants*. Macmillan & Co., 15s. net.

BOULGER, G. S., *Wood*. Arnold, 12s. 6d.

BOWER, F. O., *Plant-Life on Land*. Camb. Univ. Press, 1s. net.

DARWIN, F., *Elements of Botany*. Camb. Univ. Press, 4s. 6d.

FORBES, A. C., *English Estate Forestry*. Arnold, 12s. 6d.

—— *Development of British Forestry*. Arnold, 10s. 6d.

GILLANDERS, A. T., *Forest Entomology*. Blackwood, 15s. net.

GROOM, P., *Trees and their Life Histories*. Cassell & Co., 25s. net.

HANSON, C. O., *Forestry for Woodmen*. Clarendon Press, 5s. net.

IRVING, H., *How to know the Trees*. Cassell & Co., 3s. 6d. net.

KERNER, A., and OLIVER, F. W., *The Natural History of Plants*, 2 vols. Blackie, 30s. net.

NISBET, J., *The Forester*. Blackwood, 42s.

—— *British Forest Trees*. Blackwood, 6s.

Report of the Departmental Committee on Irish Forestry [Cd. 4027]. Wyman, price 6½d.

Report of the Departmental Committee on Forestry in Scotland [Cd. 6085]. Wyman, price 1s. 3d.

RUSKIN, J., *Modern Painters*. Dent (Everyman's Library), 5 vols., 1s. each.

SCHIMPER, A. F. W. *Plant Geography*. Clarendon Press, 42s.

SCHLICH, W., *A Manual of Forestry*. 5 vols. Bradbury, Agnew & Co., £2. 7s.

—— *Forestry in the United Kingdom*. Bradbury, Agnew & Co., 2s.

SCHWAPPACH, A., *Forestry*. J. M. Dent & Co., 1s.

Second Report (on Afforestation) of the Royal Commission, Vol. II, Part 1, 1909 [Cd. 4460]. Wyman, price 6d.

STEP, E., *Wayside and Woodland Trees*. Warne & Co., 6s.

TANSLEY, A. G., *Types of British Vegetation*. Camb. Univ. Press, 6s.

The Nature Book. Cassell & Co. 6 vols.

WARD, H. M., *Trees*. Camb. Univ. Press, 5 vols., 4s. 6d. each.

APPENDIX II

DIAGRAMS AND LANTERN SLIDES

E. J. Arnold & Son, Ltd., Leeds, and 37, Jamaica Street, Glasgow. "The A. L. pictures of trees." The 25 sheets in this series include all the well-known timber, fruit and ornamental trees. Size of each picture, 33 in. × 25 in. ; on cloth, rollers, and varnished, each 4s.

Lantern slides will be found useful for revision and for questioning. If the blinds are partly drawn down the lighting of the room can be adjusted so that it shall be possible both to see the picture on the screen and also to write or draw at desks.

The Botanical Photographs Committee of the British Association have prepared a catalogue of photographs and of lantern slides, both amateur and professional. The catalogue may be obtained from Prof. F. E. Weiss, The University, Manchester, but inquiries about slides and photographs should be addressed direct to the photographers.

The following firms have lantern slides of trees on the market :

C. Baker, 224, High Holborn, W.C.

British Botanical Association, Holgate, York. Autumn colours by Lumière process.

W. B. Crump, M.A., 4, Marlborough Avenue, Halifax. Yorkshire woodlands, &c.

G. Cussons, Ltd., The Technical Works, Broughton, Manchester. Section 37, Forest Trees ; Section 38, Forest Botany ; Section 39, Commercial Timbers.

Flatters & Garnett, Ltd., 32, Dover Street, Manchester, S.E. Section XXIII, Photo-micrographs, transverse sections of stems ; Section XXXI, Tropical vegetation.

A. Gallenkamp & Co., 19, Sun Street, Finsbury Square, London, E.C.

Henry Irving, The Rowans, Horley, Surrey. "Wayside and Woodland," a long series of British Trees, summer and winter, leaves, flowers, &c.

Newton & Co., 3, Fleet Street, Temple Bar, London, E.C. Set BO, Fruit-Tree Cultivation; Set ABQ, Amentiferae; Set 8BL, Remarkable Trees and Plants; Set VO, Botanical Slides, Photomicrographs; Set VR, Timber.

W. Watson & Sons, Ltd., 313, High Holborn, W.C. Histological Botany, transverse sections of stems, &c.

Wilson Bros., Loch-head House, Aberdeen. Set 125, Trees, &c.

APPENDIX III

MATERIAL

Fresh materials for each lesson (*e.g.* sprays of Eucalyptus and Cactus plant for ch. XIV) can be obtained by post from the British Botanical Association, Holgate, York. This firm also supplies dry material illustrating dispersion of seeds, &c., and specimens of **timber** with bark attached (price from 6*d.* each). These last can also be obtained from Messrs A. Gallenkamp & Co., 19 and 21, Sun Street, Finsbury Square, London, E.C.

Botanical pressing paper (price about 1*s.* 1*d.* per quire, size 16 in. × 20 in.) is supplied by Messrs Gallenkamp or by Messrs West, Newman & Co., 54, Hatton Garden, London, E.C., or by Messrs Watkins & Doncaster, 36, Strand, London, W.C.

Beautifully mounted **shavings** of various kinds of wood are to be obtained from Messrs Flatters and Garnett, 32, Dover Street, Manchester.

Young trees for planting can be obtained from most well-known nurserymen, although some specialize in other directions. It will probably be best to visit the nearest local nurseries, where the trees can be seen before purchasing, and to make pressing inquiries as to how often the trees have been transplanted. In case it is necessary to send to a distance the following list, which might be greatly extended, may be consulted. Several of these firms issue detailed price catalogues full of information and beautifully illustrated.

Austin & McAslan, 89, Mitchell Street, Glasgow.
J. Backhouse & Son, Ltd., The Nurseries, York.
Dickson, Ltd., Nurserymen, Eastgate Street, Chester.
A. Dickson & Sons, Ltd., Newtownards, Co. Down.
W. Fell & Co., Ltd., Nurserymen, Hexham.
The King's Acre Nurseries, Ltd., Hereford.
J. Perkins & Son, Nurserymen, Northampton.
E. Sang & Sons, Nurserymen, Kirkcaldy, Fifeshire.
J. Veitch & Sons, Ltd., Royal Exotic Nursery, Chelsea, London.
A. Waterer, Knaphill Nursery, Woking, Surrey.
Wood & Ingram, The Old Nurseries, Huntingdon.

Visits to Museums.

There is a good deal to be learnt by visiting the Museums of some public institutions, such as the Natural History Museum, Cromwell Road, S.W., the Edinburgh Museum of Science and Art, the Owens College Museum in Manchester, the Dublin Museum, and so on. Mr Henry Irving's very beautiful photographs can be seen in some of these Museums. Such a visit will be much more profitable to the class if the teacher is able to go beforehand for a private view and arrange a suitable series of questions, the answers to which can be found by searching among the specimens and their labels.

APPENDIX IV

REVISION QUESTIONS

[Mr Hugh Richardson has kindly supplied me with the following supplementary questions for revision of practical work. They may be tried as opportunity offers, not necessarily in the same months as the corresponding chapters.]

Branching and Tree forms.

1. Visit the gardener's stick heap. Collect twigs, one each, of as many different kinds as you can. The twigs may suitably be 6 to 12 inches long. Bring them indoors and wash your hands. Lay the twigs out one at a time on white paper. Sketch round the outline

with a pencil, marking the position of each bud. In this way it is easy to get the drawing the right length, but not the right thickness. Remove the twig from the paper. Use indiarubber and finish the drawing accurately in pencil. When your teacher has passed the drawing, ink it in.

If you are working by artificial light, hold the twig a few inches above the paper and parallel to it so as to cast a shadow. Try to hold it steady whilst you sketch the outline of the shadow.

2. [Homework.] Draw a family tree representing as far as you can all your relations who can be traced back to the same ancestors as yourself. Name the trunk after the original family ancestor, the main branches after his children, and so on until you come to the twigs named after your brothers, sisters and cousins.

Evergreens (Walk in Pine Wood).

1. Does the wood consist of pine trees only? Are the pines all the same sort? If more than one sort, how many kinds and what kinds? Can you distinguish them by their bark? by their leaves? by the arrangement of branches?

2. Are the trees all of about the same height or of very different heights? How high are the highest? the lowest? Do the smaller trees grow among the larger ones? or quite apart? or do the larger trees shade off into smaller ones?

3. At what distances apart are the youngest trees? Are the older trees at the same distances?

4. Is there any evidence to show whether the trees have been planted or are self-sown?

5. Are the younger trees growing so far apart that you can walk between them without touching them? or do the branches of one tree interfere with those of the next? Can you see the sky when you look up through the branches of the older trees, or do they form a complete canopy?

6. Is the floor below the trees quite bare ground? or peat, rock or sand? is it covered with pine needles or carpeted with moss or grass or herbs? or is there any undergrowth of heather, gorse, bracken or bramble? Do you find any other trees, *e.g.* birch or rhododendron, growing underneath the pines or in the more open

spaces of the pine wood ? Can you account for the special nature of the floor or undergrowth as you have seen it ?

7. Are any fir-cones lying on the ground ? Are any on the trees ? if so, on which trees and whereabouts ? Are the cones on the ground just like those on the trees ?

8. Are the lower branches of the trees alive or dead ? Are any loose branches lying on the ground ? Which kind of tree have they come from ? Have they fallen off or been cut off ?

9. Is any timber-felling going on ? if so, which kind of tree is being cut down ? What is the diameter of the trees when cut across ? Can you count the number of rings ?

10. Can you trace the roots of the trees on the surface of the ground ? How far do they extend from the trunk ? Are there any ditches or quarries where you can see how deep the roots grow ?

11. Has any damage been done to the wood by wind ? Are any trees permanently bent or lopsided ? Have any trees been uprooted ? If so, which sorts ? Is the damage on any particular side of the wood ?

12. Has any damage been done by fire ? If so, to which trees ? Have they been completely destroyed or only partly burnt or scorched ? Have the scorched trees been killed ? How do you suppose the fire started ? Could anything have been done to protect the trees ? Has anything been done since ?

13. Are any trees loaded with snow ? Can you shake the snow off the branches ? Does this alter the amount of bending ? Has any damage been done by snow ?

14. Is the soil in which the pine trees grow wet or dry, hard or soft, frost-bound or thawed ? Have any ditches been dug to drain the wood ?

15. Is any damage being done to the wood by rabbits, deer, cattle or other animals ? or by beetles, fungi, or other pests ? Are there squirrels in the wood ? Do you see any birds there, *e.g.* owl, gold-crested wren or magpie ?

16. Have any of the trees been damaged by branches being cut or broken off ? Have the wounds healed ? Has the bark grown

over them or has the wet rotted into the tree at these places ? Do you see any turpentine oozing ? from what kind of places does it come ?

17. If leave can be obtained, collect and bring home a set of specimens to illustrate what you have seen, twigs, fir-cones, moss, &c.

18. [Not to be done without leave from the owner of the wood.] Collect any branches lying on the ground and make a bonfire. Do you start the fire with bracken, pine needles or larch twigs ? Which kind of branch burns best ? What is left after the fire has burnt out ? ash or charcoal ?

Evergreens (indoors).

1. Twigs of three different pines are required. Show by drawings and descriptions how the specimens A, B, C differ from each other.

2. A bundle of sticks cut for firewood, strips cut from a soft deal box, scraps from a joiner's shop, pocket-knife. Cut at the specimens of wood provided. Try both cutting across the grain and splitting lengthwise. Describe the differences between the different woods.

3. A Christmas tree or an Araucaria in pot (price 3s. ?). Make a freehand drawing of the tree and show the arrangement of leaves, twigs, and branches. Finish in pen and ink.

Catkins (out of doors).

1. Collect twigs of alder, larch, poplar, and any other catkin bearing trees which you can find.

2. How many different sorts of catkins are there on an alder tree ? What do you understand by the large, hard, woody catkins ?

3. Is there anything lying on the ground beneath the poplar trees ?

4. Shake a hazel bush on some dry sunny day. Can you see any pollen coming off it ?

5. Next examine with a pocket lens one of the red tufted nut-forming flowers of the hazel. Can you see any tiny yellow grains on the feather tufts ?

6. What are the small rosy red cones on the larch trees ? What else do you find on the same trees ?

7. Which come first on the oak trees, the leaves or the catkins ? or are neither out yet ?

8. Do you find more than one sort of catkin on the bog myrtle ? If so, do they grow on the same bush or on different bushes ?

Thorns and Climbing plants (out of doors, to be taken in summer if possible).

1. Write a list of all the climbing plants you can find in the garden. Say for each how it climbs. Tabulate thus :

Name of plant	*How it climbs*
Pyrus japonica	bits of cloth and nails
Ampelopsis	suckers

2. Draw a full-grown tendril of cucumber which is stretching out into the air of a greenhouse. Don't touch it; but try to get the drawing exactly life-size and exactly as much curved as the object. Now stroke the tendril gently with your pencil on the concave side of its curvature. Draw it again a few minutes later.

3. Cut some sprays of white bryony, and put them quickly into warm water. If you can, avoid touching the tendrils. Repeat the cucumber experiment described above.

4. Bring some sweet pea tendrils just into contact with twigs on to which they might cling. Examine them again next day. Have they caught hold ?

5. Visit a patch of nettles, and learn all you can from them. Then write an essay "On Nettles." Read Lord Avebury's *British Flowering Plants*, pp. 355—357, on "Urtica." Go back to the nettle patch, and see what more there is to learn.

6. Collect the three most spiny holly leaves you can find and also the three least spiny. Draw all six. Can you also find intermediate patterns ?

7. Draw the thorns of blackthorn, barberry, gooseberry, and explain how they differ from each other.

8. Describe the prickles of the bramble—what shape are they? what size? on what parts of the plant are they found? How do they compare with rose prickles?

9. Walk down a hedge. Write a list of all the prickly, thorny or climbing plants you find.

Leaves (out of doors).

1. Collect one leaf—only one leaf—of each sort of tree you see during your walk. When you get indoors lay these out on a large sheet of paper. Beside each leaf write its name.

2. There will be a plant race [see *Scouting for Boys* by Lieut.-Gen. Baden-Powell, p. 104]. At the word "go" you will run round the garden, and pick just one leaf of each of twelve sorts of trees. When you get back, your teacher will give you your numbers in order; then go indoors and answer question 3.

3. Write at the top of your paper

Plant Race   [*name*]

Number in order   ——

and list of leaves brought in

1.   ——

2.   ——

3.   ——   &c.

Leaves (indoors).

4. Trace the outlines of the leaves collected in question 1. Then paint them using not only a wash of green paint, but trying also whether purple can be used to bring out the shadows.

5. Draw a single leaf in its natural position, not necessarily laid flat on paper. Use a mixture of ink and water to paint your drawing so as to show light and shade.

6. [For Homework.] Make a leaf mosaic design for embroidery. Show what colours you propose to use.

INDEX

Printed in the United States
By Bookmasters